Engineering Mechanics for Beginners

Dr.D.V. Ramamurthy

Published by

Engineering Mechanics for Beginners

978-93-86638-77-9

Author

Dr.D.V. Ramamurthy

Bonfring

309, 2nd Floor, 5th Street Extension, Gandhipuram,

Coimbatore-641 012.

Tamilnadu, India.

E-mail: info@bonfring.org

Website: www.bonfring.org

Phone: 0422 4213231

Acknowledgement

The author expresses his deep sense of gratitude to teachers for their blessings and teachings. In particular, thanks to Prof.N.L.N. Rao for the courses Theory of Machines and Operating Systems at Andhra University College of Engineering through which I learnt the fundamentals of kinematics and analytical thinking. Also, I am ever indebted to Dr. V.P. Agrawal at IIT Delhi for the mentorship and his patience with me. All that is good in these pages is but a reflection from my teachers and all the blemishes are purely from me.

Support from all my family members and blessings of my parents are the sources of energy which are taking me to places which I could not have dreamt of through my own mental capabilities alone.

Dr.D.V. Ramamurthy

Author Profile

Dwivedula Venkata Ramamurthy, graduated from Andhra University College of Engineering, Visakhapatnam with a B.E. degree in Mechanical Engineering in the year 1987, from Indian Institute of Technology, Delhi with M.Tech. in Design of Mechanical Equipment in the year 1992, and from Oklahoma State University, Stillwater, OK, USA with Ph.D. degree in Mechanical Engineering, in the year 2005. Ramamurthy worked extensively in the area of Control Systems Theory and implementation of control algorithms. Ramamurthy has to his credit, over 23 years of experience in teaching Control Systems and Theory of Machines.

References

1. Stephen Timoshenko, D. Young, Engineering Mechanics, and J. Rao, Engineering Mechanics, McGrawHill Special Indian Edition, Revised 4th Edition, 2006.
2. Vijay Kumar Reddy, K and Suresh Kumar, J. (Author), Singer's Engineering Mechanics Statics and Dynamics, B.S. Publications, 3rd Edition (SI Units), 2011.
3. Hibbeler, R.C. Engineering Mechanics Statics and Dynamics, Prentice Hall Publishers, 8th Edition, 1997.

<table>
<tr><th>Unit</th><th>Contents</th><th>Page No</th></tr>
</table>

ప్రార్థన

టిమపెంకో దేవాయ నమ:

హిట్బైలర్ దేవాయ నమః

సింగర్ దేవాయ నమః

శ్రీ విష్ణు ప్రకాశాయనమః

నరసింహ దేవతాభ్యం శిరసా నమామి

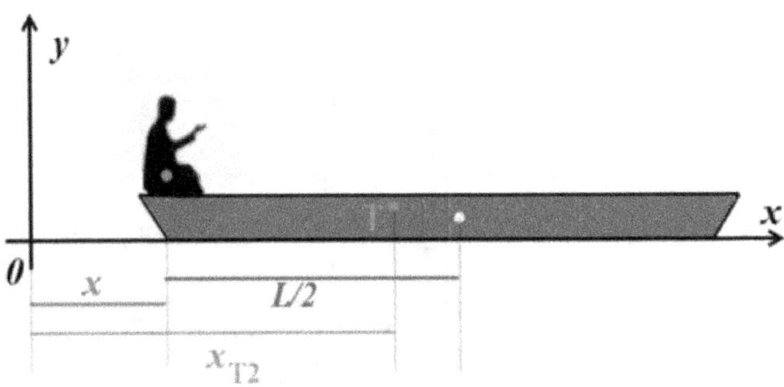

ENGINEERING MECHANICS FOR BEGINNERS

1. Why this Book?

There are many books on Engineering Mechanics, and excellent ones too. These books deal with Engineering Mechanics in a beautiful way, presenting the subject in depth. The presentation sometimes is very lengthy and the so called "present-day-student" doesn't find time to cull through all the details. However, if the student is already familiar with the jargon, he'd be able to digest the gist. The idea behind this book is exactly that: to stand as an *appetizer* or a *starter* before the student goes on to the divine main *course* presented by the authors invoked in the prayer. In fact, as compared to those books, this book may pale into insignificance ! I strongly urge all students of engineering mechanics to buy at least one of the three books: by Hibbeler, by F.L. Singer, or by Timoshenko.

My guess would be this book can be "studied" in less than 20 hours and the student can do that before delving into the course. And then, this book would come handy as a ready revision material towards the end-semester examinations. Also, it may prove its usefulness to the busy "modern-day-teacher who needs to quickly glance through the main points before stepping into the class-room.

The book is meant to be a picture-book of Engineering Mechanics giving importance to intuitive feel and the graphical appeal to the student. There are about 300 figures and illustrations spread through the book to make the reading more of visual scanning than framing ideas by combining words. Most of the book is devoted to statics and only very fundamental part of dynamics is presented.

If you enjoyed the book please let me know. If you didn't enjoy the book at all, then please do let me know too ! My email address is ramdwivedula@gmail.com

1.1. Engineering Mechanics

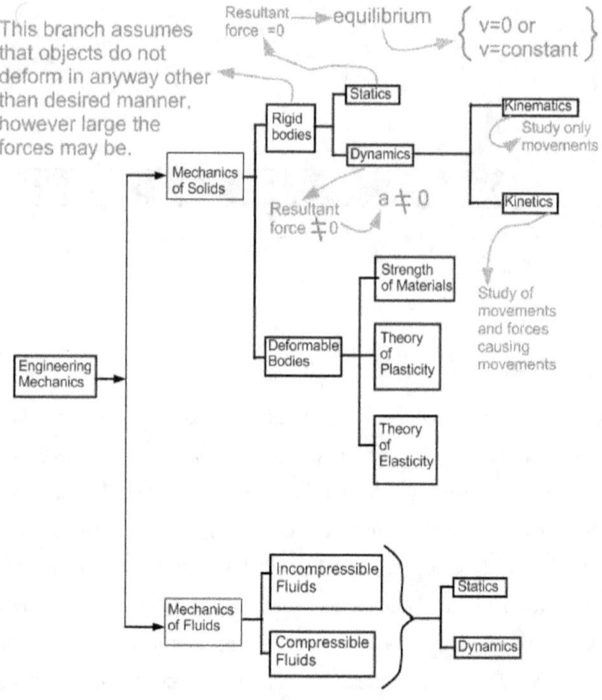

Mechanics can be defined as that branch of the physical sciences concerned with the state of rest or motion of bodies that are subjected to the action of forces.

Engineering refers to the application of sciences to solve practical problems with application of scientific results with the aim of improving human comfort and productivity. Engineering involves design and fabrication of devices to provide comfort to humans. Thus, "Engineering Mechanics" refers to the application of knowledge of forces, their effects, and movements to design and fabricate devices/contrivances with the aim of improving comfort and improving efficiency of human beings.

The classification scheme shown above is only representative and slightly different versions may appear in different textbooks. This shouldn't bother us in the course we are studying because we are interested only in the boxes which are annotated with textual description associated with them. We begin the study of engineering mechanics with study of axioms.

1.1.1. *Axioms*

Axioms are self-evident facts which cannot be proven mathematically, in the sense of the word "proof", but can only be demonstrated to be true using evidences. All the "Theorems" can be "proven" using axioms by mathematical techniques of proof; essentially this involves demonstrating that the "theorem" to be proven is equivalent to one or more of "axioms". There are seven axioms that form the basis of Engineering Mechanics, together with lot of engineering "common sense[1]."

First Axiom, Newton's First Law

Everybody[2] continues in its state of rest or of uniform motion unless compelled by an external agency.

Newton's first law defines force as that external agency which changes the state of a body. Also, we see that Newton's first law **defines equilibrium** as state of rest or of uniform motion.

Second Axiom, Newton's Second Law

The time rate of momentum is directly proportional to the external force and acts in the direction of the force.

Newton's second law lays foundation of dynamics. The "time rate of momentum" indicates derivative, $\frac{d}{dt}$. Thus, Newton's law may be stated as

$$F \propto \frac{d}{dt}(mv) \implies F = k\frac{d}{dt}(mv)$$

Where k is the constant of proportionality. Further, if mass of the object is constant, it can be taken out of the derivative. Hence,

$$F = km\frac{d}{dt}v = kma.$$

In SI units, 1 N of force (i.e., F=1), acting on 1 kg of mass (i.e., m=1), produces $1\ \frac{m}{s^2}$ of acceleration (i.e., a=1). Substituting these values into the last equation gives 1=k×1×1 or k=1. Thus,

$$F = ma.$$

It may be noted that "F=ma" is not statement of Newton's Second Law.

Third Axiom, Newton's Third Law

Action and reaction forces are equal, opposite, and are collinear.

---

[1]Which is perhaps very "uncommon"
[2]One may read it as "Everybody" to see newer insight.

When two objects are in contact, each exerts a forces on the other. Force exerted by any one object ON the other may be taken as "action force" and the force exerted BY the other object may be taken as "reaction force.".

It is implied here that both the objects are in equilibrium.

Fourth Axiom: Newton's Law of Gravitation

This law is often expressed in equation form as

$$F = G\frac{m_1 m_2}{r^2}$$

where F is the force of attraction between two objects, G is the universal constant of gravitation equal to $66.73 \times 10^{-12}\frac{m^3}{kg-sec^2}$, m_1, m_2 are masses of the objects, and r is the distance between the Centers of Gravity[3] of the objects.

All the objects on earth are attracted by earth. Let m_2 be the mass of earth, r be the radius of earth.

Then the force with which the earth attracts the object having mass of m_1, computed according to Newton's Law of gravitation as

$$F = G\frac{m_1 m_2}{r^2} = m_1\frac{G m_2}{r^2} = m_1 g.$$

With the value of G = $66.73 \times 10^{-12}\frac{m^3}{kg-sec^2}$, $m_2 = 5.972 \times 10^{24}$ kg , and r = 6371 km the value of g comes out to be $g = 9.81\frac{m}{s^2}$. This is referred to as ***acceleration due to gravity***. Though theoretically value of g varies from point to point on earthy very slightly, it is taken to be constant.

The quantity $m_1 g$ represents the force with which earth attracts an object having mass m_1 and is called the "weight of the object on earth."

The next three axioms, which relate to "statics", predate Newton's laws by centuries. However, Newton's laws are generally stated as the first four axioms because of their importance. The next three axioms require the definition of "Resultant."

Definition: Resultant of a system of forces action on an object is that single force[4] which produces the same effect as the system of forces.

---

[3]This term will be defined more precisely later.
[4]Remember this definition carefully but also remember that in some situations it may not be possible to reduce the system of forces to a single resultant force.

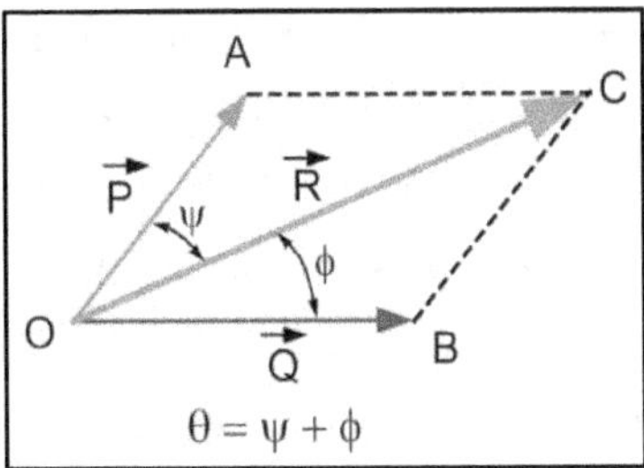

Definition: *A set of forces is said to be in equilibrium if the magnitude of their resultant is zero.*

Fifth Axiom, Parallelogram Law of Forces

Parallelogram law of forces states that the resultant of two forces acting on an object is obtained as the diagonal of the parallelogram formed with two forces as adjacent sides.

Figure beside shows parallelogram law. Two vectors, $\vec{P}$ and $\vec{Q}$ meeting at a point and subtending an angle θ are acting on an object. The resultant of these two forces is shown by the vector $\vec{R}$. This vector, as per the Parallelogram law, is shown as the diagonal of the parallelogram OACB formed from the two sides OA and OB. The length OA is scaled proportional to the magnitude of $\vec{P}$ and the length OB is scaled the same proportional to the magnitude of $\vec{Q}$. When using graphical method, we draw the parallelogram and measure the length of OC and multiply the same scaling used for representing OA and OB. Parallelogram's law may be represented in equation form as:

$$R = \sqrt{P^2 + Q^2 + 2PQ\cos\theta} \ \text{ with } \phi = \sin^{-1}\left(\frac{Q}{R}\sin\theta\right) \text{ and } \psi = \sin^{-1}\left(\frac{P}{R}\sin\theta\right).$$

Sixth Axiom, Superposition Principle[5]

The state of an object does not change even if a set of forces in equilibrium is either added or subtracted.

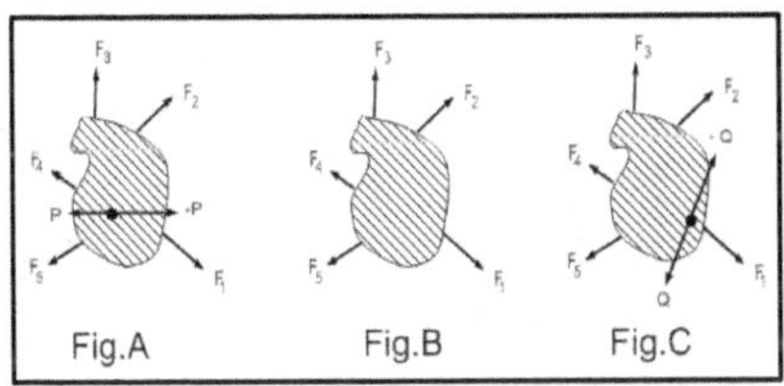

---

[5]This principle, in a sense is equal to the mathematical equation a+0 = a for all a.

According to this axiom, the state of the object shown hatched is the same in the situation shown in Fig.A, or in situation shown in Fig.B in which the sub-system of forces P and - P are subtracted, or in Fig.C in which a sub-system of forces Q and - Q are added.

Seventh Axiom, Law of Transmissibility of Force

The external effect[6] of a force on an object does not change even when the point of application of the force is moved to a new location along the line of action of the force.

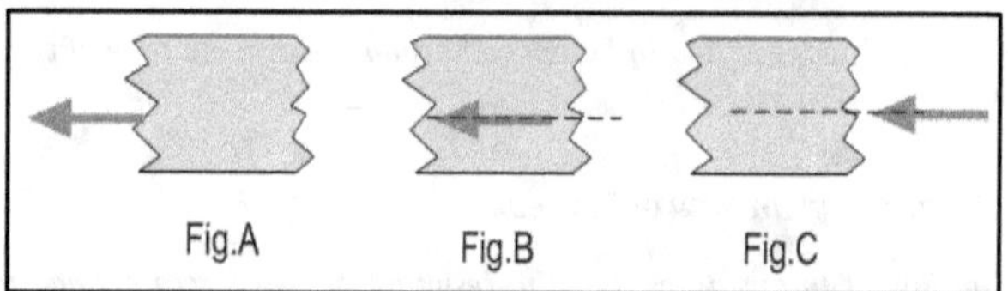

It is intuitive that whether we *apply* the force either at the surface of the object or whether we apply the force through a string at a distance, the state of the object doesn't change. Thus, the external state of the object in Figures A, B, and C is the same according to the Law of Transmissibility of force.

Many students consider that "Engineering Mechanics" is a "tough" course. But really, if we understand how to apply these seven axioms and apply our[7] "common sense", one can consider that one has mastered Engineering Mechanics and all its allied subjects.

1.2.   Free-Body Diagram

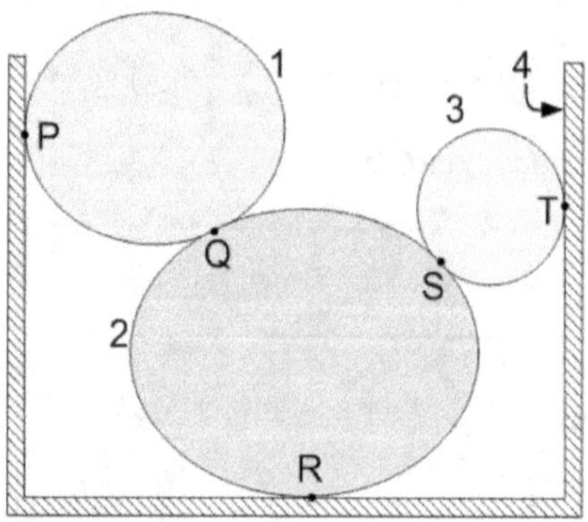

---

[6]Note carefully that only "external" effect doesn't change. "internal" effect may change. Think about it.

[7]Once again, we note that "common sense" is very "uncommon." In fact if one develops very strong "common sense", that person will go a long way in the career and life!!

S.No.	Object	Forces acting on the object	Number of forces acting on the object
1	Sphere 2	✓ Contact force at Q ✓ Contact force at R ✓ Contact force at R ✓ Weight of Sphere 2 ✓ Force F_1	4
2	Sphere 3	✓ Contact force at S ✓ Contact force at T ✓ Weight of Sphere 3	3
3	Sphere 1	✓ Contact force at P ✓ Contact force at Q ✓ Weight of Sphere 1	3
4	Container 4	✓ Contact force at T ✓ Contact force at R ✓ Contact force at P ✓ Weight of container	4

Real-life systems are built from individual objects and/or subsystems which are in contact[8]. Very often, it will be of interest to the engineer to compute all or certain forces acting on certain bodies either to analyze the system or use the information in design of the system. In such situations, the engineer would want to focus the attention only on the specific object and/or subsystem and represent all the forces acting on that object. Typically, these forces are represented in a diagram and this diagram is called "free-body diagram", abbreviated as FBD. As an example, consider three spheres, 1, 2, and 3, put in a container as shown in Figure. Let us say that we are interested in finding the force with which the sphere 2 will press onto the container (shown as object 4 in the figure). Since there are four distinct objects in this case, four FBDs can be drawn: one FBD for each object. Once we identify this, we may list forces acting on each of the objects in a table as shown. It is a good practice to tabulate the forces the way shown, at least mentally, so that no force is left unaccounted. Notice that the contact force Q is listed as a force both on Sphere 1 and on Sphere 2. Similarly all contact forces are shown to act on both the bodies in contact. This is quite natural because when two bodies are in contact, each body exerts a contact force on the other as per Newton's third law.

Once the forces are identified, we will try to draw the FBD of each of the objects. The question now is : Where should be begin? It is a good practice to begin with that object which is acted upon by least number of forces.

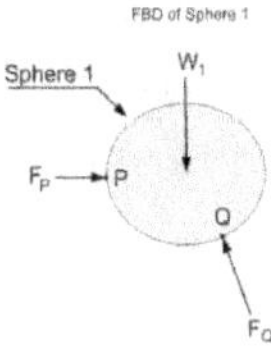

---

[8]Or which exert force without coming in contact such as magnetic force.

So we may begin with either Sphere 1 or Sphere 3. Let us choose Sphere 1 and draw its FBD as shown in Figure. In the FBD, we will show only the object considered and the forces acting on that object. There are three forces, weight of the Sphere[9] , contact force[10] at P, and contact force at Q. No other objects are shown on FBD of Sphere 1. There are three forces in the FBD and typically, at least one force is specified in the problem as a known force. Lami's Theorem[11] may be used to compute the two unknown forces in this case.

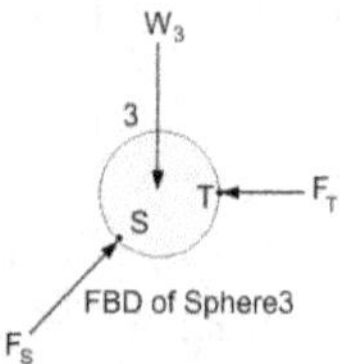

Once we are done with the FBD of Sphere 1, we will look into the table and identify the next object which is acted upon by least number of forces. We see that it is sphere 3 and its free body diagram is drawn. as shown in Figure beside. Again, we see that there are three forces and Lami's Theorem may be used to compute the unknown forces. Notice that F_S is alone the line joining the centers of Spheres 3 and 2.

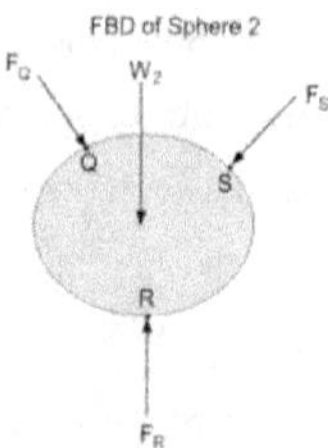

Again, we refer to table and pick the next object for drawing the free body diagram. There are two objects left: Sphere 2 and the container. Let us choose Sphere 2 and draw its FBD. The trick of applying Axiom 3 given on page 3 is shown in the FBD of Sphere 2 and carefully look at the directions of the forces in the FBD of Sphere 2. The force F_Q shown in FBD of Sphere 2 is the force exerted by the Sphere 1 on Sphere 2 whereas the force F_Q shown in FBD of Sphere 1 is the force exerted by the Sphere 2 on the Sphere 1. Similarly, F_S shown in FBD of Sphere 2 is the force exerted by Sphere 3 on Sphere 2. Since the number of forces is more than three, we use method of projections listed on page 18 to compute the unknown forces.

---

[9]Weight is always chosen to act vertically down and is shown to act at the center of gravity of the object.
[10]The contact force acts normal to the contact surfaces. In this case, the common normal passes through the centers of the spheres 1 and 2.
[11]This will be discussed later. Don't worry about it now.

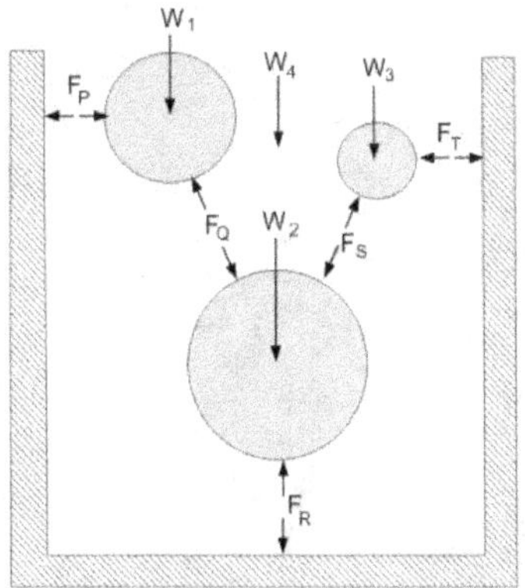

FBD of Container,4

The only object which is left for drawing the FBD is the container 4 which is shown in Figure beside. Carefully notice the directions of all the forces. Notice that the force representing the weight of the container, W_4, is not touching the container. It appears untrue but it is very much true as see in the discussion on centroid and Center of gravity presented on page 37 onward.

All the forces shown on the FBDs shown so far may be "assembled" and shown on one single diagram. This diagram is shown in next page. Such view, when drawn in machine drawing, is sometimes called "exploded view."

There are five contact points in this case viz., P, Q, R, S, and T. At each point, there is a pair of forces which are equal, oppositely directed, and collinear. If you do not understand this fact and the FBDs of each of the objects thoroughly, do not proceed further. If may confuse you and obfuscate all further understanding.

To summarize, the procedure for solving problems involving several bodies may be specified as:

(i) Count the number of objects in the problem. Very many times, "ground" or fixed object is not counted.

(ii) Count the number of forces on each object.

(iii) Start with that object for which the number of forces is least.

- Draw the object
- Identify points where the object is in contact with other objects
- At each of these points of contact, show a contact force
- Show the weight of the object, if it is not to be ignored.
- Determine unknown forces using equations of equilibrium.

(iv) Repeat step (iii) until all objects are considered.

With the free body diagram drawn, the next task usually is to find unknown forces using any of the axioms. Typically this process involves using any of the graphical methods available or invoking the equations of equilibrium. Such topics are considered in subsequent pages.

1.3.   Reaction Forces

When two bodies touch each other, each one exerts a force on the other. If A and B are two objects, force exerted by A on B may be termed "action force" and the force exerted by B on A may be termed "reaction force." It is essential to know the nature of this contact force to be able to draw Free body diagrams. A few types of contacts and the reaction forces generated are illustrated here using line sketches.

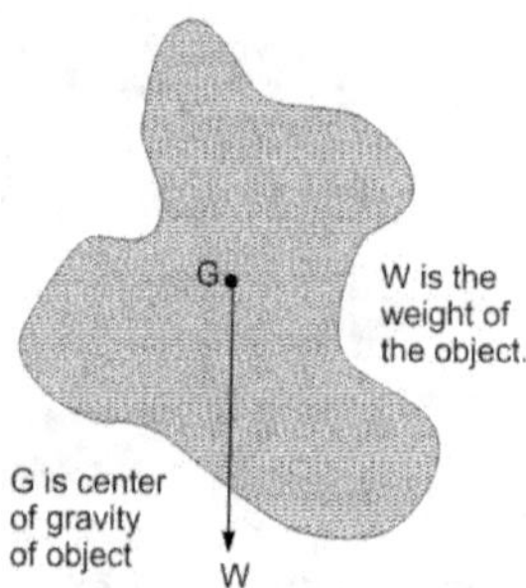

Gravity Force

According to Newton's law of gravitation (see page 4), every object attracts every other object. So every object on earth attracts earth and consequently earth attracts every other object. The force with which earth attracts an object is termed "weight of the object." The line of action of this force is directed from the center of gravity[12] of the object to the center of earth. By convention, gravity force acting on a body is always shown downwards.

Wire/Rope/Cable/Spring Force

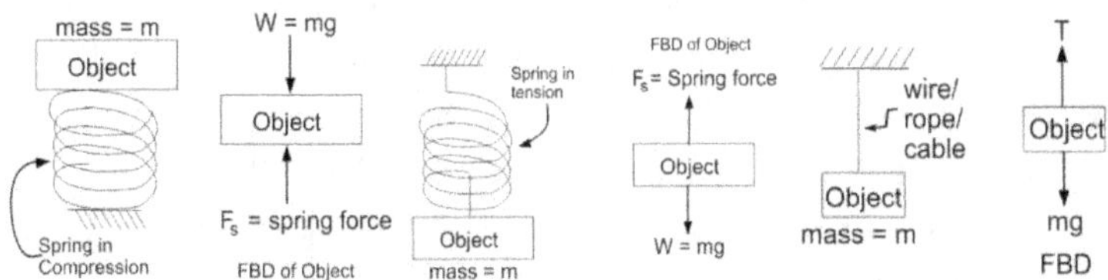

Flexible elements such as wire/rope/cable exert force on an object hanging from their end along the direction of their axis. Further, they exert force only when they are taut. Springs, which are also flexible members, can exert force when they are in tension or in compression. In the figure shown beside, notice that the "Object" is pulling the wire down and the wire is pulling the object up. Hence the force T shown in the FBD of the object may be considered as the reaction of the wire. When objects are "connected" to springs, springs exert forces too. Springs may be in tension or in compression as shown in Figures. When spring is not "loaded" (or not in contact with any external object), its length is called "free length". When a spring is in tension, its length increases. When the spring is in compression, its length decreases. If we represent a vector displacement as

$$\vec{x} = \left\{ \begin{array}{c} \text{position of} \\ \text{end of spring} \\ \text{when loaded} \end{array} \right\} - \left\{ \begin{array}{c} \text{position of} \\ \text{end of spring} \\ \text{when not loaded} \end{array} \right\} = \text{loaded length} - \text{free length},$$

then the force exerted by the spring on the object, as shown in the FBD of object will be $F_s = -k\vec{x}$ where k is the spring constant and k has units N/m or N/mm. Thus, we see that the direction of "reaction" due to a spring is opposite to the elongation or compression.

Smooth surface

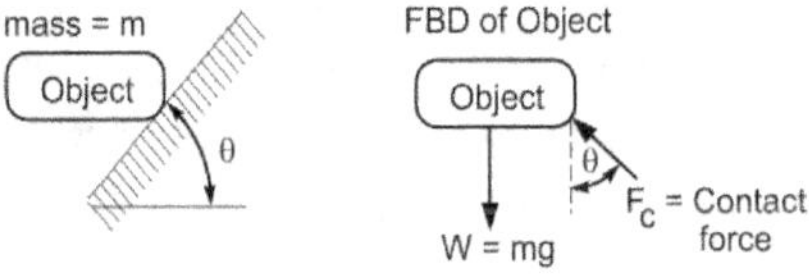

When an object touches the smooth[13] surface of the other object, the contact force will be normal to the surface. Carefully notice the angle θ in the schematic and the FBD. There will be

---

[12]Center of gravity will be discussed later.

as many contact forces as the number of points where the object touches other objects. Further, notice that if the contact force F_c is zero , then the contact at that point is either lost completely or is about to be lost.

As a further example, consider the case of two circular objects touching each other as shown. Roller A is touching two other objects: (i) Wall at point P and (ii) Roller B at Q. So there will be two contact forces F_P and F_Q as shown in the FBD of Roller A. Further, if m_1 is the mass of Roller A, a gravity force m_1g needs to be shown in the FBD of Roller A. Similarly three forces F_Q , m_2g, and F_R act on Roller B as shown in the FBD of Roller B. Convince yourself that the forces F_P, m_1g, and F_Q meet at the axis of the Roller A in its FBD and that the forces F_Q, m_2g, and F_R meet at axis of Roller B in its FBD.

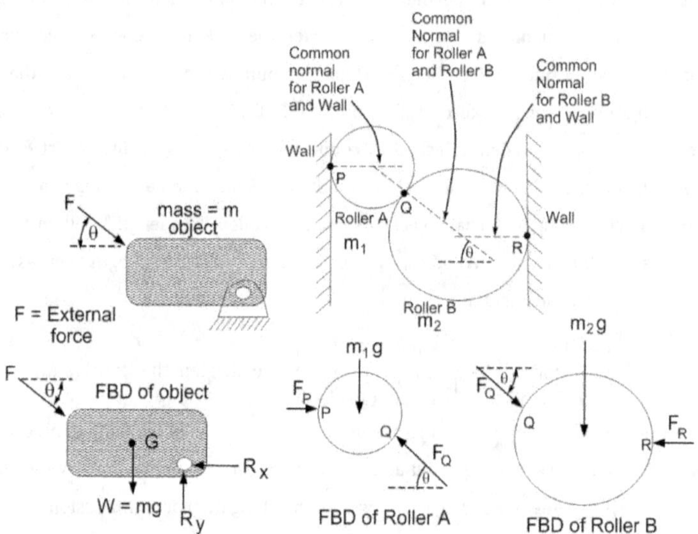

Hinge or Pin-Connection

In some situations, objects are put into contact and are constrained to have only rotary relative motion. Such objects are said to be connected by "hinge" or "pin connection." A hinged joint is often shown as indicated in the figure.

There are two components of the hinge or pin reaction as shown[14] in the FBD of Object viz., R_x and R_y.

---

[13]Friction force existing between contact surfaces is ignored here. For discussion on dry friction, refer to page30.
[14]Check carefully whether the directions of R_x and R_y are correct here.

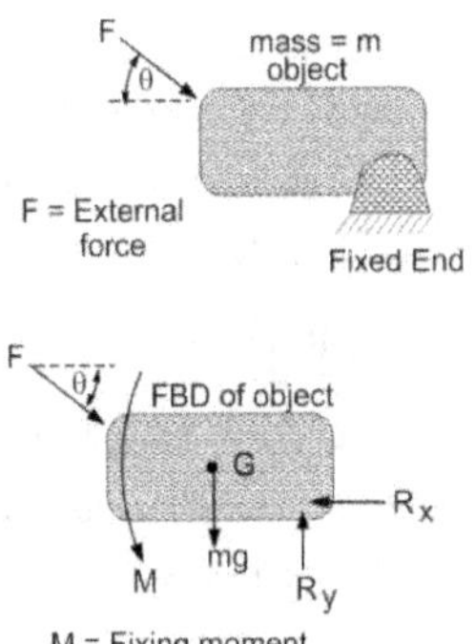

Fixed Connection

When one body is connected to the other by a fixed connection, no relative movement is possible at the fixed connection. A fixed connection offers two reaction forces R_x and R_y and a fixing moment M as shown in the figure. We need to notice that the fixing moment and the reaction forces are so adjusted that they resist the external loading. If there is no external loading at all then there will be no reactions at all. The same applies to the hinged connections also.

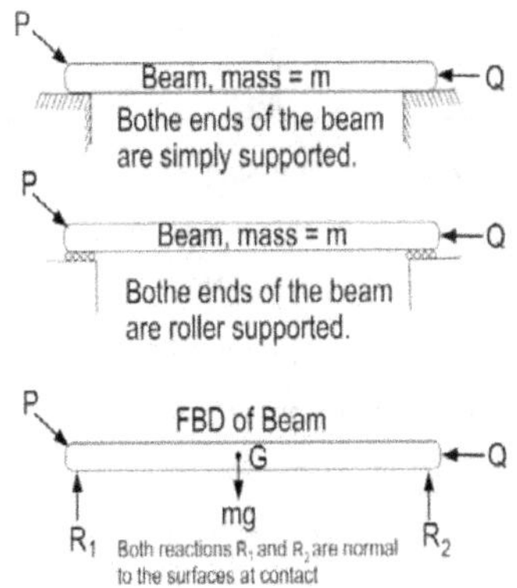

Simply Supported or Roller Support

If the connection is maintained solely by the weight of the object being supported, such connection or support is called "simply supported." Sometimes when very smooth contact is needed, roller(s) may be placed in between the objects in contact to reduce the friction. The reaction force for a simply supported or roller supported case is normal to the contact surfaces as in the case of smooth surface.

1.4.   Resultant of Forces

When a set of forces act on a body, the state of the body changes according to the magnitudes and directions of the forces. This module begins to consider the process of finding the "net effect" of all the forces acting on a body.

We study this in four situations as listed below:

1.   Resultant of Coplanar, Concurrent Forces acting on a body.
2.   Equilibrium of a system of coplanar, concurrent forces.
3.   Resultant of coplanar, non-concurrent forces
4.   Resultant of a system of parallel forces
5.   Equilibrium of a system of parallel forces.

Resultant of Coplanar, Concurrent Forces

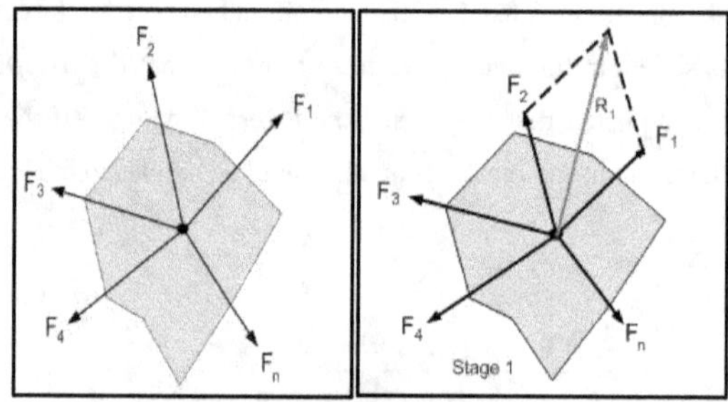

Coplanar forces are forces which act in a single plane and concurrent forces are forces which meet at a point. A set of coplanar, concurrent forces is shown in the figure beside in which "n" number of forces are acting on a body. Notice that we have shown the forces as F_1 , F_2 , ..., F_n rather than $\vec{F}_1, \vec{F}_2, \cdots, \vec{F}_n$. The arrow on the head of the symbol to denote vector quantity is omitted when we understand from context that a vector quantity is at hand. The resultant of this system of forces, R is given by

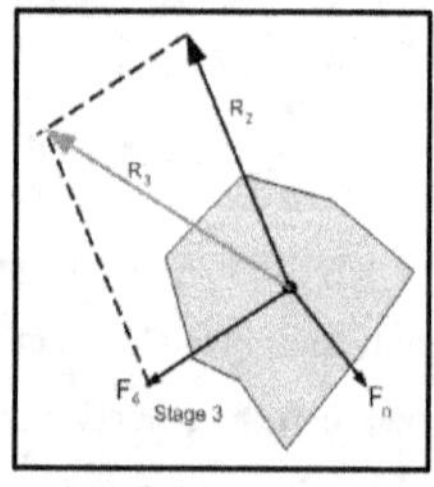

$$R = F_1 + F_2 + \cdots + F_n.$$

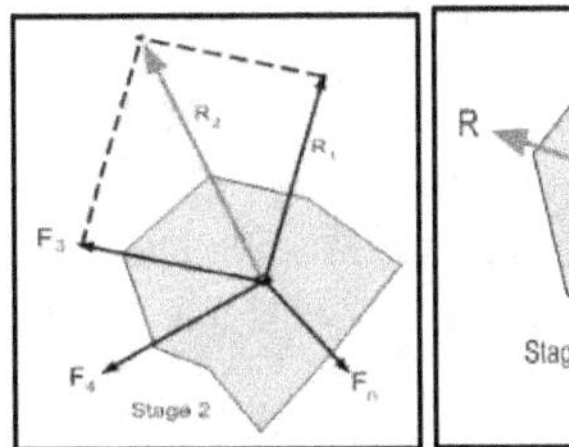

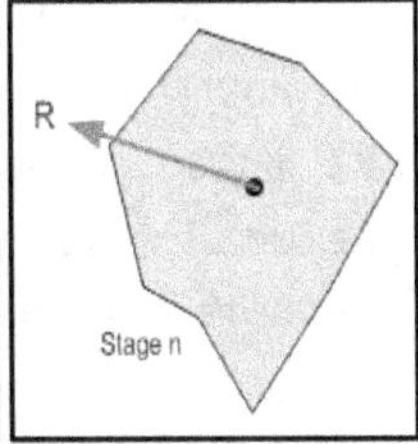

The resultant is found by applying Parallelogram law successively. In stage 1, F_1 and F_2 are added to get R_1 as shown in stage 1 figure beside. Having obtained R_1 , F_1 and F_2 may be removed R_1 may be put in their place. Now, the resultant of R_1 and F_3 may be found as shown in Stage 2 figure. R_1 and F_3 may now be removed from the picture and R_2 may be included as shown in stage 3 picture. This process may be applied consecutively till only one resultant remains as shown in stage n figure. This single force is the resultant of the system of forces. It may be noted that a system of coplanar, concurrent forces always has a resultant force.

Polygon Law for Finding the Resultant of Coplanar, Concurrent Forces

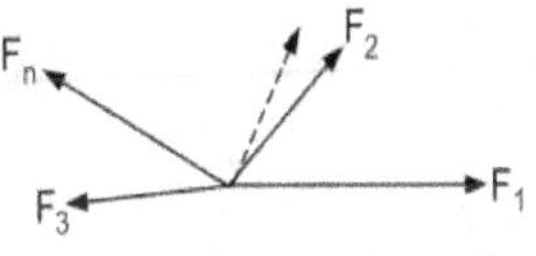

Figure(a): Coplanar, Concurrent Forces

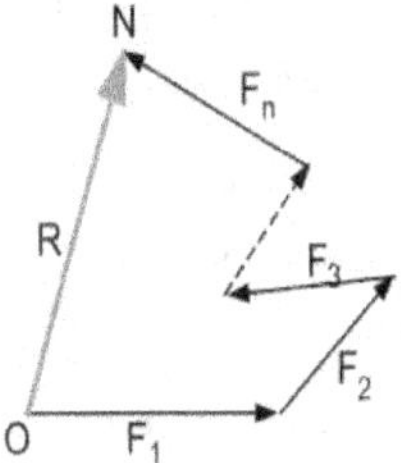

Figure(b): Force Polygon for resultant

The resultant of a set of coplanar, concurrent forces is given by the closing side of the polygon drawn by placing the force vectors from head to tail.

This idea is shown in Figure beside where F_1, F_2, ..., F_n are the vectors representing the coplanar, concurrent forces and R is their resultant. The forces whose resultant needs to be found is shown in Figure (a). To find the resultant, first draw one of the forces, say F_1 to some scale starting from point O. From the head of F_1, draw the vector F_2; from the head of F_2, draw the vector F_3 and so on. When all the force vectors are shown in the force polygon shown in Figure (b), join O and N by directed line segment *going* from O to N: this represents the resultant of forces. <u>If, after drawing the force polygon, we find that the points O and N coincide, that the length of vector ON is zero, we say that the forces are in equilibrium.</u> That is if the resultant is of zero length, the forces are in equilibrium. This fact is stated as "Polygon Law for equilibrium."

Polygon Law for Equilibrium

If a set of forces are in equilibrium, then the force polygon drawn by placing vectors representing the forces from head to tail will be a closed polygon.

In Figure (a), set of forces are shown. In Figure (b), the force polygon is shown starting from point O. If the head of the last vector, F_n, coincides with the point O, with which we began, we say that the system of forces is in equilibrium

Two theorems regarding coplanar, concurrent forces are stated here without proof. Readers are encouraged to go through the proofs given in any of the standard text books.

Three Forces in a Plane Theorem

If three forces keep a body in equilibrium and two of them meet at a point, then the third force must lie in the plane formed by the two forces and must pass through the point of intersection of the first two forces.

This theorem finds it use in identifying the direction and/or line of action of one of the unknown forces. Reader is encouraged to go through the proof of this theorem.

Lami's Theorem

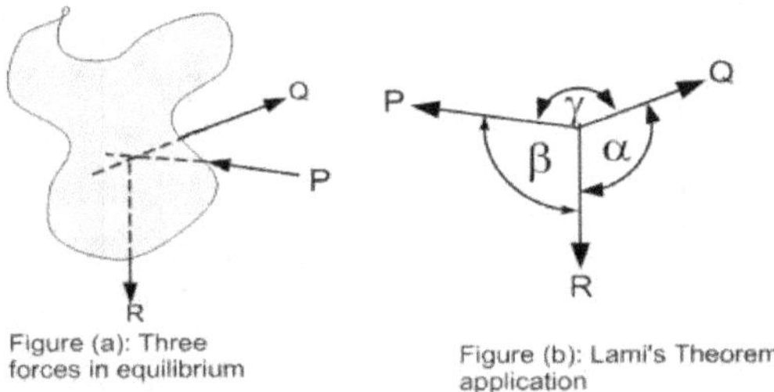

Figure (a): Three
forces in equilibrium

Figure (b): Lami's Theorem
application

If three coplanar concurrent forces keep an object in equilibrium, then magnitude of each force is proportional to the sine of angle between the other two. Further, the constant of proportionality is the same for all three forces.

Lam's Theorem in equation form may be written as:

$$\frac{P}{\sin \alpha} = \frac{Q}{\sin \beta} = \frac{R}{\sin \gamma}$$

Applying parallelogram law successively takes a lot of time especially if the number of forces is more. Method of projections is very convenient approach for finding the resultant. Let us first define what a projection is

Projection

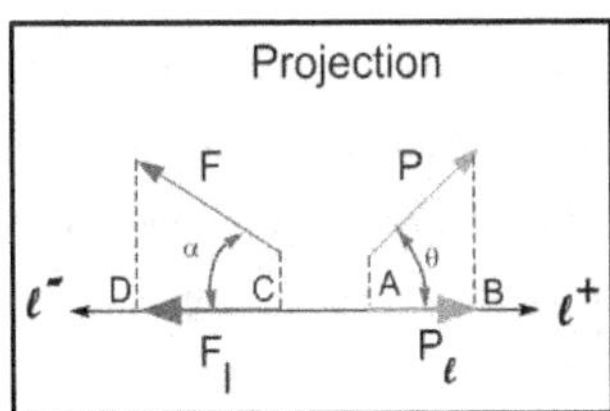

Consider the line l which has positive and negative directions, l+ and l- associated with it as shown in Figure. Consider force P shown and drop perpendiculars from the tail and head to meet the line l at A and B. The force represented by line going from A to B is called the Projection of the force P onto the line l, P_l. Since P_l is along l+, we say that P_l is positive and write $P_l = +P \cos \theta$. Similarly if we consider a force F and drop perpendiculars from the tail and head onto the line l to meet it in C and D, the force represented by line going from C to D is called the projection of F onto l, F_l. Since the direction of F_l is along l-, we say that F_l is negative and write $F_l = -F \cos \theta$.

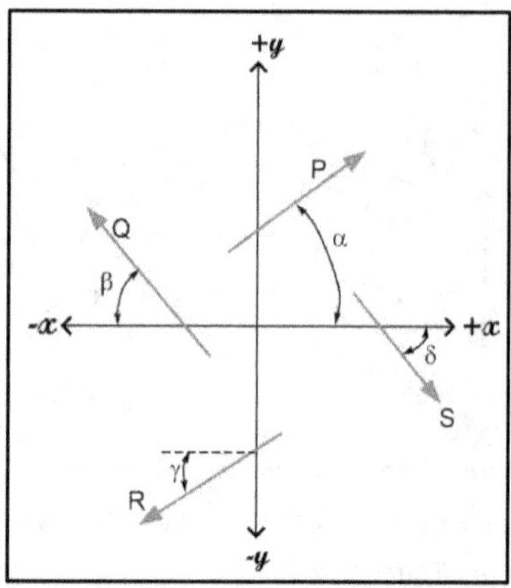

When solving problems, we often setup coordinate axes x and y and take projections on to them. Consider four forces shown in Figure beside. The projections of these forces onto x and y axes may be shown in Table below.

Force	x-component	y-component	Quadrant
P	$P \cos \alpha$	$P \sin \alpha$	First
Q	$-Q \cos \beta$	$Q \sin \beta$	Second
R	$-R \cos \gamma$	$-R \sin \gamma$	Third
S	$S \cos \delta$	$-S \sin \delta$	Fourth

Notice that the angles α, β, γ, and δ are all angles made by the forces with x axis as shown and all are acute angles. Note this very carefully. Also, notice that in First quadrant, both x and y components are positive; in Second quadrant, x- component is negative but y component is positive; in Third quadrant both x and y components are negative; finally in fourth quadrant, x component is positive but y component is negative.

These facts need to be borne in mind because we use these all through the course. Now, we consider the problem of finding the resultant of a system of coplanar, concurrent forces and the problem of establishing the condition of equilibrium of a set of forces.

Method of Projections: To Find the Resultant of a Set of Forces

Let F_1, F_2, F_3, ..., F_n be n forces acting on a body. To find the resultant of these forces, find the sum of the x-components of these forces, and sum of y-components of these forces.

$$R_x = F_{1x} + F_{2x} + \cdots + F_{nx} = \sum_{i=1}^{n} F_{ix},$$

$$R_y = F_{1y} + F_{2y} + \cdots + F_{ny} = \sum_{i=1}^{n} F_{iy}$$

where F_{1x}, F_{2x}, ..., F_{nx} are the x components of the forces F_1, F_2, F_3,..., F_n and F_{1y}, F_{2y}, ..., F_{ny} are the y components of the forces F_1, F_2, F_3,..., F_n . Depending on the signs of R_x and R_y we can decide in which quadrant the resultant lies e.g., if $R_x>0$ and $R_y<0$, the resultant lies in fourth quadrant etc. The magnitude of the resultant is found by[15]

$$R = \sqrt{R_x^2 + R_y^2}$$

and the angle R makes with x axis may be found by

$$\theta = \tan^{-1}\left(\frac{|R_y|}{|R_x|}\right).$$

Condition of Equilibrium of Coplanar, Concurrent Forces

Let F_1, F_2, F_3, ..., F_n be n forces acting on a body. These forces will be in equilibrium if

$$R_x = F_{1x} + F_{2x} + \cdots + F_{nx} = \sum_{i=1}^{n} F_{ix} = 0 \text{ and}$$

$$R_y = F_{1y} + F_{2y} + \cdots + F_{ny} = \sum_{i=1}^{n} F_{iy} = 0.$$

Notice that while using method of projections, we need to take care of the signs of the components.

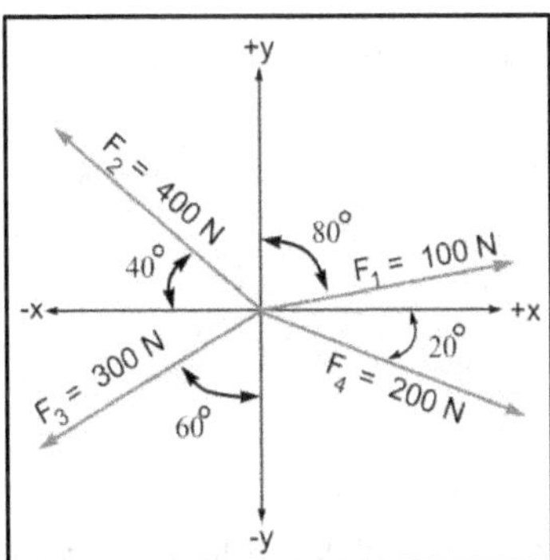

1. *Example 1: Resultant of Coplanar Concurrent Forces*

Find the resultant of the force system shown in Figure.

---

[15]Does this result come from Parallelogram law?

Solution

Sum up the x and y components of the forces

$$R_x = 100\cos 10° - 400\cos 40° - 300\cos 30° + 200\cos 20°$$
$$R_y = 100\sin 10° + 400\sin 40° - 300\sin 30° + 200\sin 20°$$

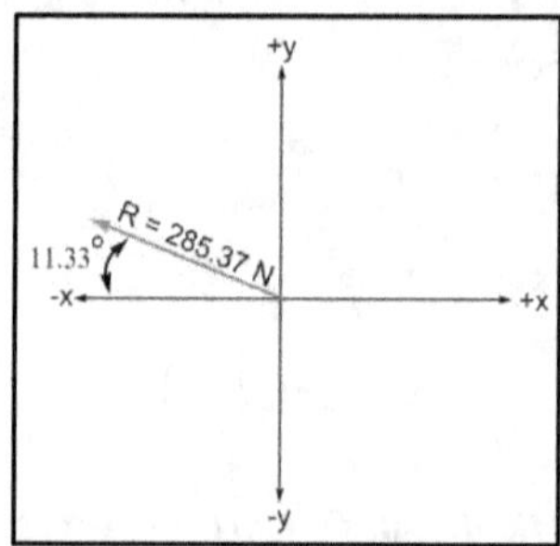

Or, R_x = -279.8 N and R_y=56.07 N. Since R_x is negative and R_y is positive, R lies in second quadrant.

Magnitude of $R = \sqrt{R_x^2 + R_y^2} = \sqrt{279.8^2 + 56.07^2} = 285.37\ N$ and $\theta = \tan^{-1}\left(\frac{|R_y|}{|R_x|}\right) =$ 11.33°. The resultant force, R, is shown in Figure beside. By the definition of resultant, this single force produces the same effect on an object as the four forces F_1, F_2, F_3, and F_4 produce.

2. Example 2: Three Cylinders in a Rectangular Ditch

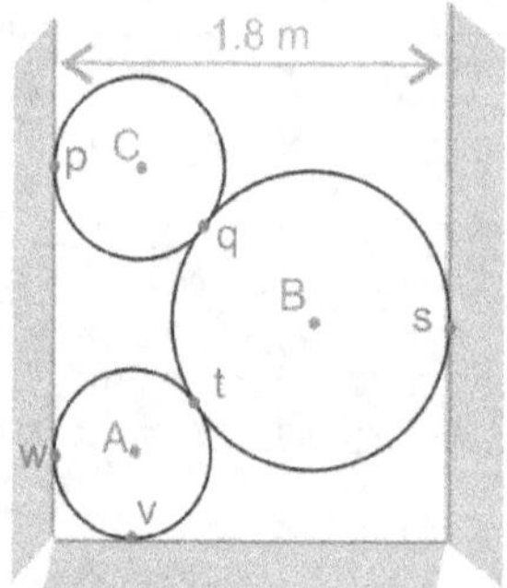

Three cylinders, A, B and C are piled in a rectangular ditch as shown in Figure beside. $W_A = 300N$, $r_A = 0.4m$, $W_B = 800N$, $r_B = 0.6\ m$, $W_C = 400N$, $r_C = 0.5m$. neglecting friction, determine all contact forces. $[R_p = 330\ N;\ R_q = 518N;\ R_s = 1529\ N;\ R_t = 2000N;\ R_v = 1500N;\ R_w = 1600N]$

<u>Solution</u>: Notice that the contact force between roller C and roller B will be along the line joining C and B. Similarly, the contact force between the roller A and roller B will be along the line joining A and B.

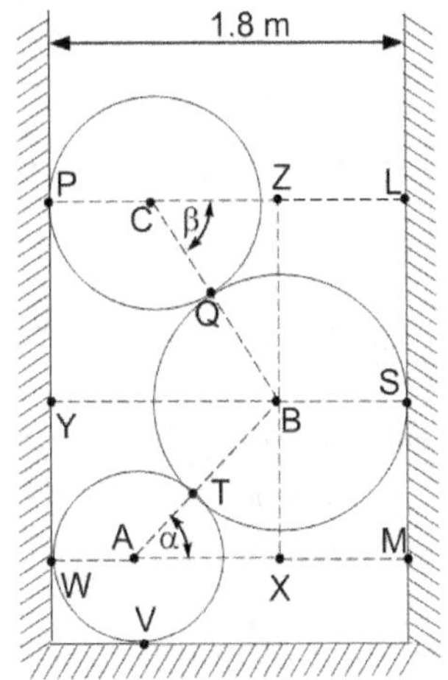

We need the inclinations of these forces with respect to the horizontal. We compute these angles before we proceed to the Free body diagrams. It is always a good idea to compute the required geometric details beforehand so that we don't have to stop for geometric details at a later stage of drawing FBD and establishing equilibrium conditions.

To that end, we draw the diagram as shown beside with all details. We need angles α and β. Notice that

$$\overset{r_A}{\overset{\frown}{WA}} + AX + \overset{r_B}{\overset{\frown}{XM}} = 1.8$$
$$\Rightarrow r_A + AX + r_B = 1.8$$
$$\Rightarrow AX = 1.8 - r_A - r_B$$
$$\text{Or, } AX = 0.8 \ m$$

and

$$\overset{r_C}{\overset{\frown}{PC}} + CZ + \overset{r_B}{\overset{\frown}{ZL}} = 1.8$$
$$\Rightarrow r_C + CZ + r_B = 1.8$$
$$\Rightarrow CZ = 1.8 - r_C - r_B$$
$$\text{Or, } CZ = 0.7 \ m$$

In triangle $\triangle ABX$, $\cos\alpha = \dfrac{AX}{AB} = \dfrac{AX}{\underset{r_A}{\underbrace{AT}} + \underset{r_B}{\underbrace{TB}}} = \dfrac{0.8}{1} = 0.8. \Rightarrow \sin\alpha = 0.6.$

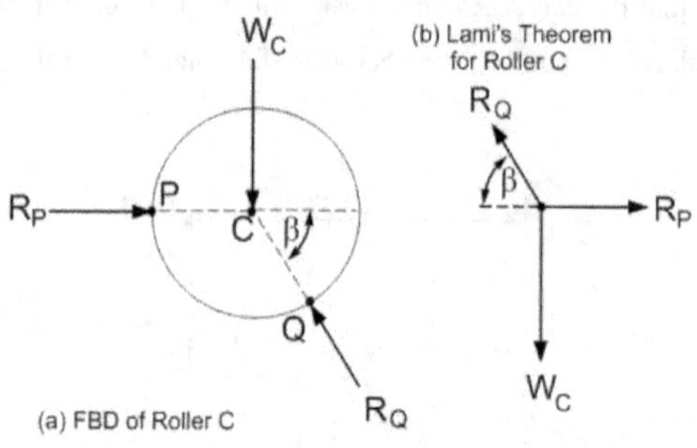

In triangle $\triangle BCZ$, $\cos\beta = \dfrac{CZ}{CB} = \dfrac{CZ}{\underbrace{CQ}_{r_C}+\underbrace{QB}_{r_B}} = \dfrac{0.7}{1.1} = \dfrac{7}{11} \Rightarrow \sin\beta = \dfrac{\sqrt{72}}{11}$.

We now begin by drawing FBD of roller C[16]. Figure (a) shows the FBD of Roller C and Figure (b) shows the application of Lami's theorem, as stated on page 17. Since Roller C is in equilibrium, we say, from Lami's theorem, that

$$\frac{R_P}{\sin(90+\beta)} = \frac{R_Q}{\sin 90^\circ} = \frac{W_C}{\sin(180^\circ-\beta)}$$

from which we write

$$R_P = \frac{W_C}{\sin\beta}\cos\beta = \frac{400}{(\sqrt{72}/11)}\frac{7}{11} \approx 330\ N,$$

$$R_Q = \frac{400}{(\sqrt{72}/11)}\sin 90^\circ = \frac{400}{(\sqrt{72}/11)} = 518\ N.$$

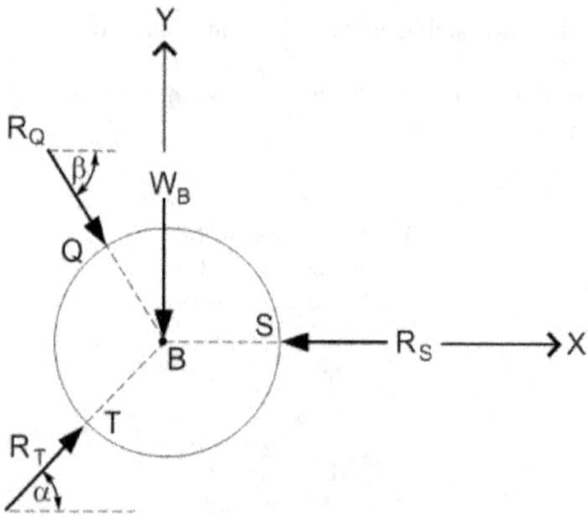

Now we draw the FBD of roller B as shown in the figure beside. There are four forces acting on Roller B. Hence Lami's theorem cannot be used. Establishing x and y axes as shown in Figure, we write equations of equilibrium[17] (see page 19)

---

[16]Why roller C?

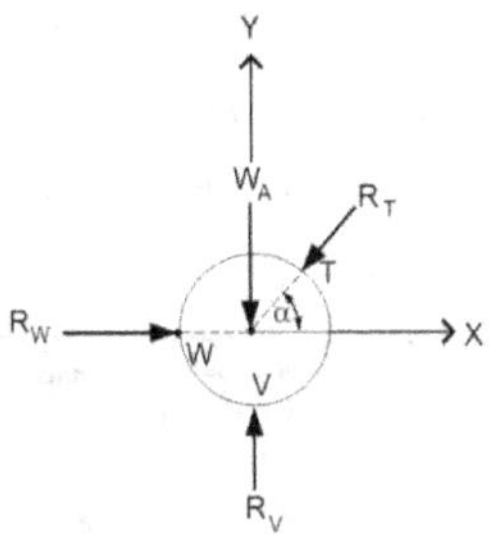

$$\sum F_Y = 0 \Rightarrow -W_B - R_Q \sin\beta + R_T \sin\alpha = 0 \Rightarrow R_T = \frac{(W_B + R_Q \sin\beta)}{\sin\alpha} = 2000\ N$$

$$\sum F_x = 0 \Rightarrow +R_Q \cos\beta - R_S + R_T \cos\alpha = 0 \Rightarrow R_S = R_Q \cos\beta + R_T \cos\alpha = 1529\ N$$

Now we draw the FBD of roller A. Establishing the xy-axes as shown, the equations of equilibrium may be written as:

$$\sum F_X = 0 \Rightarrow R_W = R_T \cos\alpha = 2000 \times 0.8 = 1600\ N$$

$$\sum F_Y = 0 \Rightarrow R_V = W_A + R_T \sin\alpha = 300 + 2000 \times 0.6 = 1500\ N$$

[It is highly advisable to go through each step of this problem very carefully and then solve the problem without looking at the solution to ensure that each and every step is understood.]

3. Exercise

Find the magnitude and direction of F_5 if F_1, F_2, F_3, F_4, and F_5 are in equilibrium.

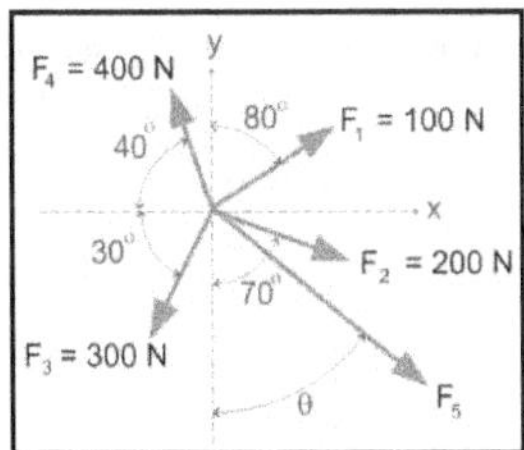

1.5.   Coplanar, Non-Concurrent Forces: Moment of a Force

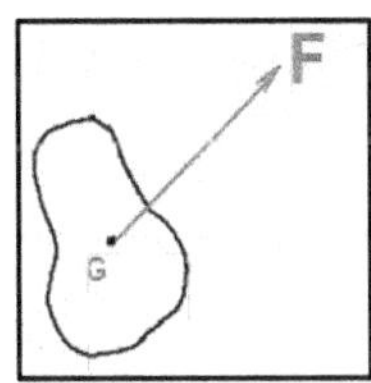

---

[17] Why did we write the equation in y-direction first?

When the line of action of the resultant force passes through the center of gravity of the object on which the force is acting, the object experiences translatory[18] motion as shown in Figure beside. The acceleration of the body will be in the direction of the force as per Newton's Second Law.

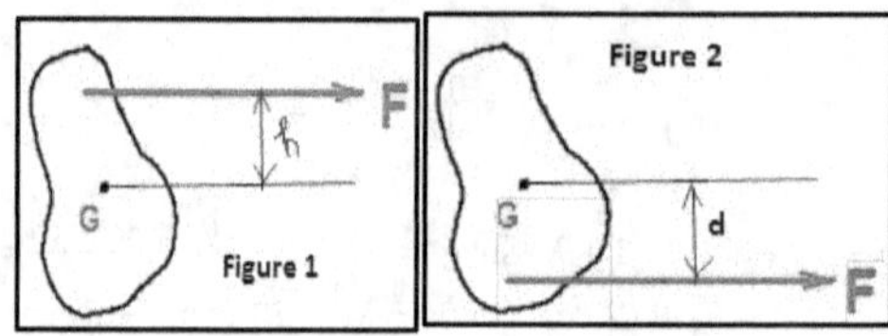

However, when the line of action of the force doesn't pass through the center of gravity of the object as shown in Figure 1, the body experiences both translatory as well as rotary movements. It is our experience that the distance "h" is an important parameter to be specified if we need to determine the effect of force on the object. Put together, F and h define a moment, the tendency of force to cause rotation: we write M=F*h. Also, we know that when Force F is applied as shown in Figure 1, the body experiences a rotary motion in clockwise direction looking from the top. However, when the force is applied as shown in Figure 2, the body experiences a rotary motion in clockwise direction looking from top. The moment created in Figure 1 is said to be negative and the moment in Figure 2 is said to be positive. For Figure 1, we say that $M_1 = -Fh$ and for Figure 2, we say $M_2 = Fd$.

When a body is in equilibrium under the action of a set of forces, the object shall neither experience a tendency for translatory motion or rotary motion. The two conditions $\sum F_x = 0$ and $\sum F_y = 0$ are already seen to be two conditions for equilibrium. With the new concept of moment, it can be seen that the three conditions for equilibrium are

$$\sum F_x = 0$$
$$\sum F_y = 0$$
$$\sum M_A = 0$$

Where x and y are **any** coordinate axes chosen conveniently and A is **any** point in the plane of the moment.

Summary: The moment of a force about a point is defined as the product of magnitude of the force and the perpendicular distance from the point to the line of action of the force. The

---

[18]A body is said to undergo Translatory motion when every point on the body undergoes same displacement as the body moves.

problem of resultant of a set of coplanar, non-concurrent forces and the problem of establishing equilibrium of such forces will be considered after an example is presented.

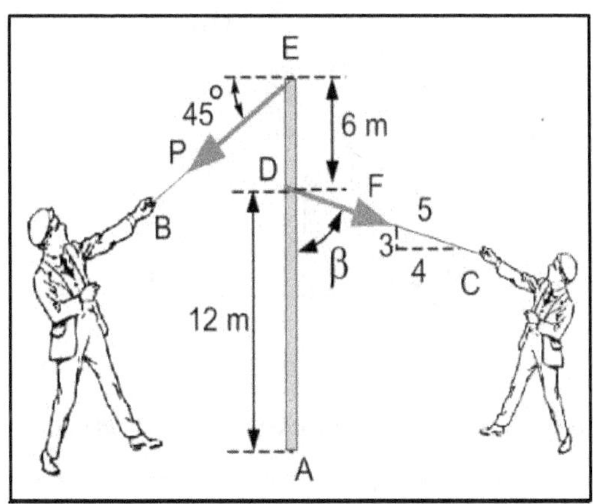

1.5.1. Example: Moment of a Force

Two men apply forces P=100 N and F = 70 N as shown in Figure below. Find the moment of each force about point A. Which way will the pole rotate, clockwise or counter clockwise?

Solution: The perpendicular distance from A to the line of action of force $P = \frac{18}{\sqrt{2}} = 12.73\ m$ and from A to that of force $F = 12 \times \frac{4}{5} = 9.6\ m.$

Thus,

M_{PA}=Moment of force P about point A = P * 12.73 = 1273 N-m (Counter clockwise) and

M_{FA} = Moment of force F about point A = -70*9.6 =- 672 N-m. (Clockwise)

M_A = The net moment about point A= M_{PA} + MFA = 1273-672 = 601 N-m.

Since the moment is positive, the pole will rotate counterclockwise about point A.

Exercise: If the force P is increased to and we want to prevent the rotation of the pole, what force F must be applied?

1.5.2. Resultant of Non-Concurrent, Coplanar Forces: Varignon's Principle

To find the resultant of a system of non-concurrent, coplanar forces, we use Varignon's principle[19]:

---

[19]Is Varignon's principle a theorem or is it an axiom? Make sure you understand this and study the proof.

The moment of resultant of a set of forces is equal to the algebraic sum of moments of each force in the set about the same point.

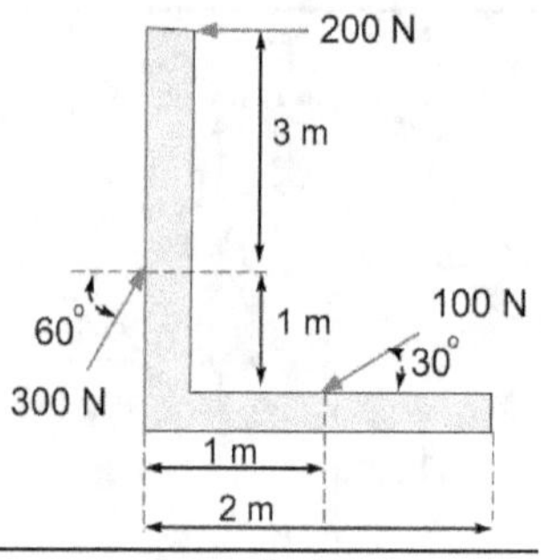

1.5.3. Example: Resultant of Coplanar Non-Concurrent Forces

Find the resultant of the forces acting on the L-Beam shown.

<u>Solution</u>: There are three details to find when computing resultant:

- Its magnitude,
- Its direction (angle), and
- Its point of application.

The magnitude and direction are found using familiar method and the point of application is found using Varignon's Principle.

To Find Magnitude

$$R_x = -200 + 300 \cos 60 - 100 \cos 30 = -136.6\ N$$

$$R_y = 300 \sin 60 - 100 \sin 30 = 209.8\ N$$

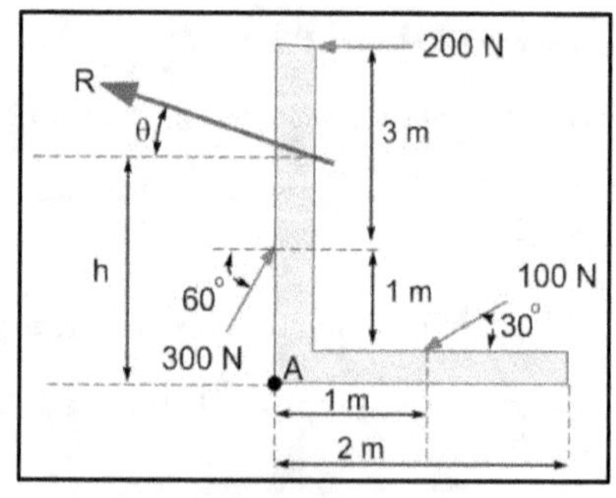

$$\therefore R = \sqrt{R_x^2 + R_y^2} = 250.4\ N$$

Since Rx<0 and Ry>0, R must lie in second quadrant as shown.

<u>To find direction</u>: $\theta = \tan^{-1}\left(\frac{|R_y|}{|R_x|}\right) = 56.9\,^\circ$

<u>To find point of application</u>: We use Varignon's Principle. Indicate the direction of R by drawing directed line segment to represent the line of action of R as shown in Figure and use the equation.

Moment of R about A = Sum of moments of forces about A to get[20]

$$(R\cos\theta \times h) = (200 \times 4) - (100 \times \sin 30 \times 1) - (300\cos 60 \times 1) = 600\,N - m$$

$$\Rightarrow h = \frac{600}{R\cos\theta} = 4.38\,m$$

Exercise: Is it possible to make $\theta = 50°$ either by increasing or by decreasing the magnitude of 200 N force?

1.5.4. *Vector Notation for Moments*

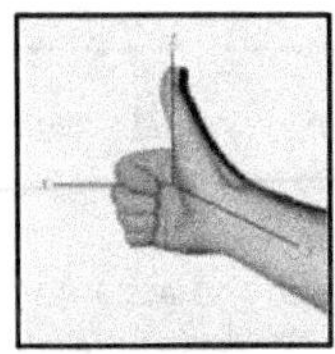

Vector notation greatly simplifies the problems in three dimensions. Several simple concepts, when understood, allow us to use vector algebra for solving problems of engineering mechanics. The first concept to learn is "Right Handed Coordinate System."

A rectangular cartesian coordinate system is said to be RIGHT HANDED provided the thumb of the right hand points in the direction of the positive z-axis when the right-hand fingers are curled about this axis and are directed from positive x towards positive y-axis as shown in Figure.

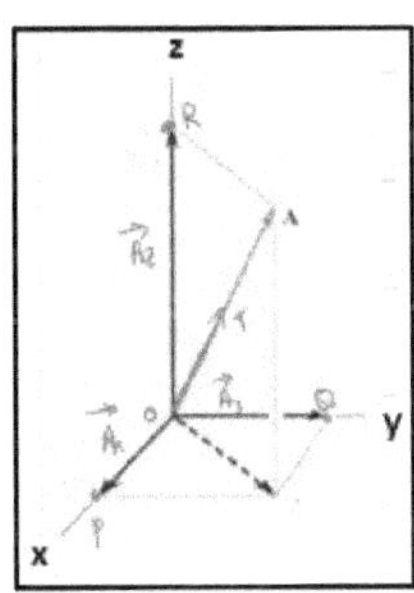

---

[20]Very carefully follow the following equation.

A few important notions, which find their use in solving engineering mechanics problems are listed below:

- Three components of a vector $\vec{A}$ may be written as $\vec{A} = \vec{A}_x + \vec{A}_y + \vec{A}_z$ as shown in Figure.

- Magnitude of $\overrightarrow{A_x} = A_x = |\overrightarrow{A_x}| = length\ of\ OP \Rightarrow A_x = OP$. Similarly $|\overrightarrow{A_y}| = A_y = OQ$.

- Magnitude of A = OA = $\sqrt{A_x^2 + A_y^2 + A_z^2}$.

- Unit force vector[21] along the direction of A $= \dfrac{\vec{A}}{|\vec{A}|}$.

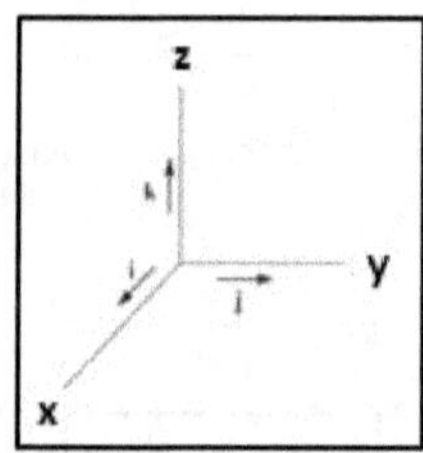

- i, j, k represent unit vectors along x, y , and z axes as shown in Figure.

- $\vec{A} = A_x i + A_y j + A_z k$

- α = Angle between x-axis and line of action of vector A. Similarly β, and γ may be defined as angles made with y and z axes.

- $\cos \alpha = \dfrac{A_x}{A}, \cos \beta = \dfrac{A_y}{A}, and\ \cos \gamma = \dfrac{A_z}{A}$ are called direction cosines of vector A.

- $\cos^2 \alpha + \cos^2 \beta + \cos^2 \gamma = 1$

- Position vector of P = $\vec{P} = p_x i + p_y j + p_z k$ where p_x, p_y, and p_z are coordinates of point P.

- $\vec{u}_p$ = Unit vector along $\vec{P} = \dfrac{\vec{P}}{|\vec{P}|} = \dfrac{p_x}{p} i + \dfrac{p_y}{p} j + \dfrac{p_z}{p} k$ where $p = \sqrt{p_x^2 + p_y^2 + p_z^2}$.

One illustrative problem is presented here.

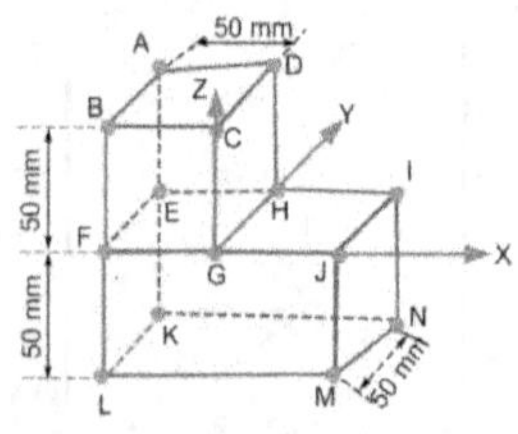

---

[21] We get a unit vector along any given vector by dividing that vector with its magnitude.

Problem: Vector Notation

The figure shown beside is made by joining two boxes together.

1. Find how many vertices/corners are there in the object.
2. How many "plane faces" does this object have. Count only outward faces.
3. How many of these faces are parallel to each other?
4. Considering x,y, and z axes shown, where is origin?
5. If i,j,k are unit vectors along x,y,z, find the position vectors of points M and K.
6. A vector goes from I to L. Find the unit vector along this line.
7. A force of 100 N magnitude $\vec{F}$ acts along the line IL. Express this force in terms of i,j,k.

Solution

1. There are twelve vertices: A, B, C, D, G, H, I, J, K, L, M, and N. E and F are not vertices.
2. There are ten plane faces.
3. BCFG, ADHE, EINK, FJML are parallel to each other. ABLK, CDGH, MNIJ are parallel to each other. ABCD, GHIJ, KLMN are parallel to each other
4. The origin, point of intersection of x,y,z axes, is located at point G.
5. Coordinates of M are (50,0,-50). Coordinates of K are (-50,50,-50). Therefore $\vec{M}$=50 i- 50 k. And $\vec{K}$ = -50 i + 50 j - 50 k.
6. $\vec{I} = (50\,i + 50\,j)$ and $\vec{L} = (-50\,i - 50\,k)$. Therefore $\vec{P} = \vec{IL} = \vec{L} - \vec{I} = -100\,i - 50\,j - 50\,k.$ Therefore , $P = \sqrt{100^2 + 50^2 + 50^2} = 122.5\;mm.$ $\Rightarrow \vec{u_p} = \dfrac{\vec{P}}{P} = -0.816i - 0.408j - 0.408\,k.$
7. $\vec{F} = 100\vec{u_p} = -81.6\,i - 40.8\,j - 40.8\,k.$

Make yourself very familiar with this vector notation. It will come in very handy, beyond imagination.

Moment in Vector Notation

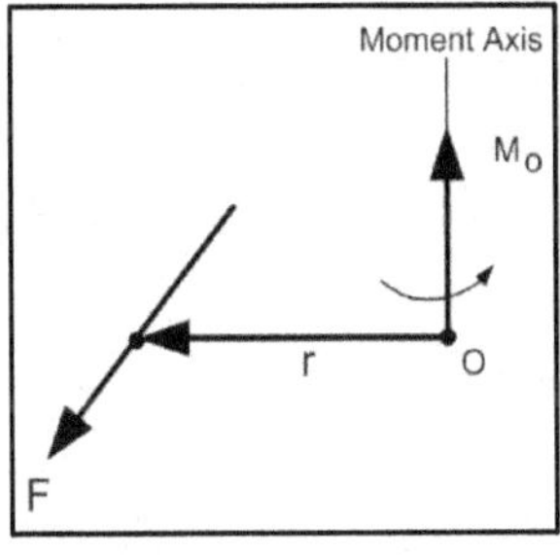

Consider a force F acting on a body as shown in the Figure beside. The moment of this force about point O (strictly speaking moment of force F about moment axis) may be written as[22]

$$M_o = r \times F$$

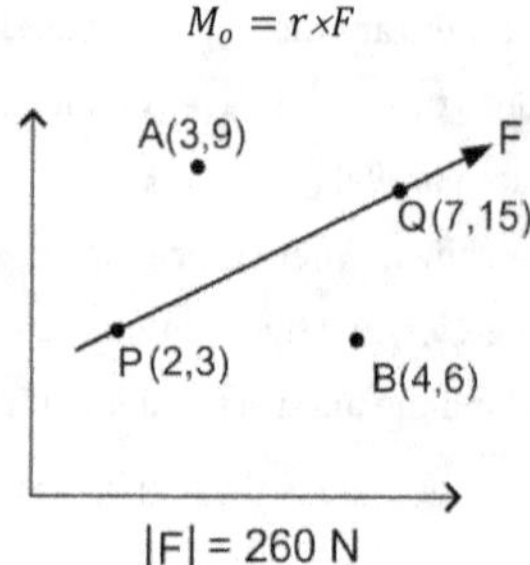

Where r is position vector going **from O to any point on the line of action of F**.

Example: Moment of Vector

Find the moment of force **F** of magnitude 260 N, about points A and B.

Solution

$$\vec{PQ} = \vec{Q} - \vec{P} = (7i + 15j) - (2i + 3j) = 5i + 12j$$

Therefore $\left|\vec{PQ}\right| = \sqrt{5^2 + 12^2} = 13$ and u_{PQ} = unit vector along PQ $= \dfrac{\vec{PQ}}{\left|\vec{PQ}\right|} = \dfrac{5}{13}i + \dfrac{12}{13}j$.

So the force vector, F may be written as, $F = |F|u_{PQ} = 260\left(\dfrac{5}{13}i + \dfrac{12}{13}j\right) = 100i + 240j$ N.

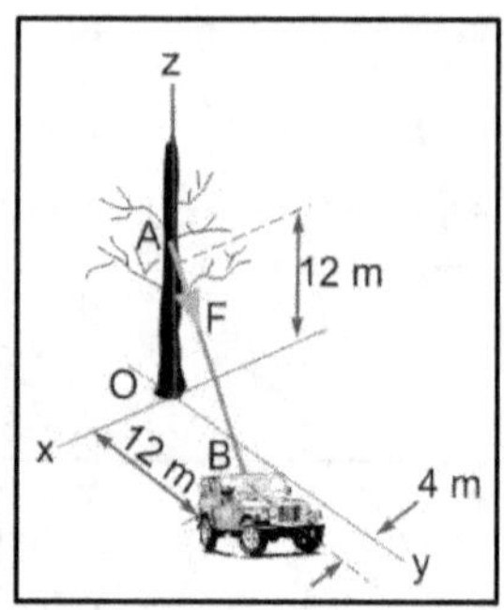

To take moment about point A, find $r_{AP} = P - A = -(i + 6j)$ and write moment as[23]

$$M_A = r_{AP} \times F = -(i + 6j) \times (100i + 240j) = 360k.$$

Similarly, $r_{BP} = (3i + 9j)$ and $M_B = r_{BP} \times F = -180k$.

---

[22]The arrows on the head of r, F, and M_o are removed since we understand that all these are vectors.
[23]Notice the sign of M_A: it is positive. Whereas for M_B it is negative. What do you infer from these?

<u>Exercise</u>: In an attempt to fell a tree, a rope is tied to point A on the tree and the other end B is tied to a jeep as shown in the Figure. If the force F exerted by the jeep is 200 N, find the moment of force F about point O.

Couple

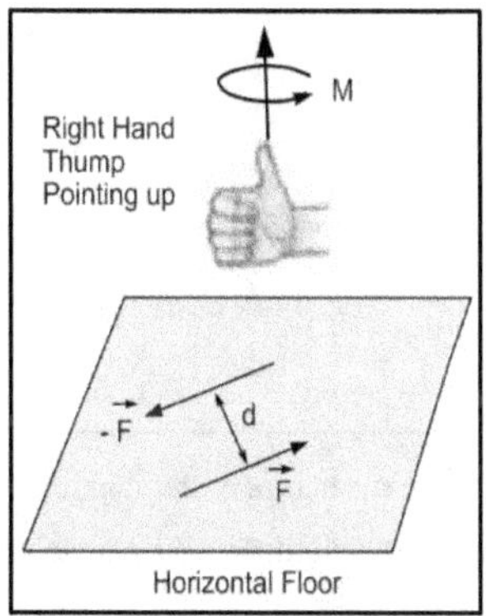

A couple is defined as two parallel, opposite forces that have same magnitude but are separated by a perpendicular distance.

An example of couple is shown in Figure beside. Note that the resultant of the two forces F and -F is zero. However, these two forces constitute a couple whose sole effect is to produce rotation or a tendency of rotation.

The moment produced by these two forces is called a "Couple." If the moment produced is counterclockwise, it is positive and negative otherwise.

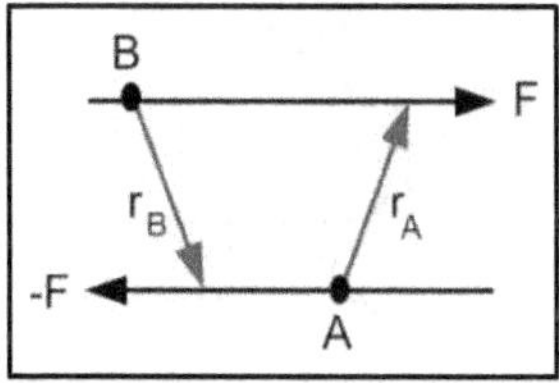

The magnitude of a couple is obtained as product of magnitude of one of the forces and the perpendicular distance between the forces.

In the figure shown below, the couple of the forces can be computed either by taking moments about point A or point B and we can write

$$M_A = M_B = r_A \times F = r_B \times (-F).$$

Example: Couple of Forces

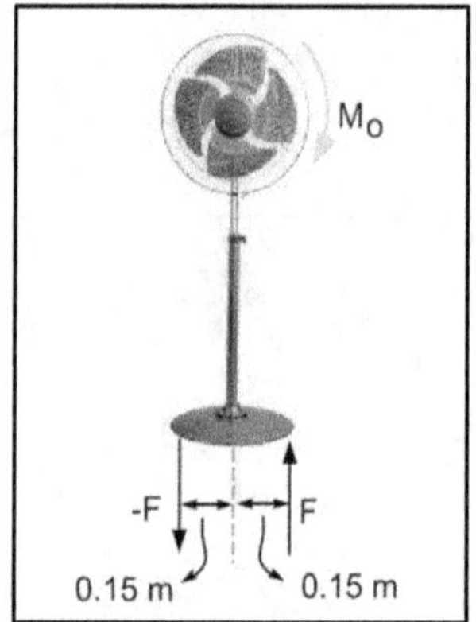

The frictional effects of the air on the blades of the standing fan creates a couple moment of 6 N-m on the blades. Determine the magnitude of the couple at the base of the fan so that the resultant couple moment on the fan is zero.

Solution: We may write

(Moment due to air friction)+(Couple due to reaction forces)=0

$$\therefore -M_o + F(0.3) = 0 \Rightarrow F = \frac{6}{0.3} = 20\ N$$

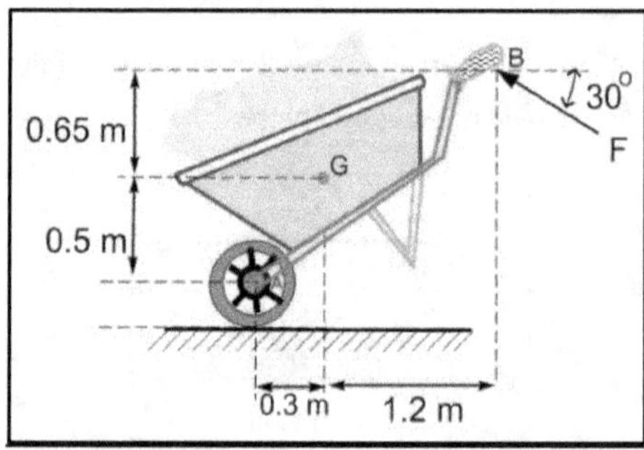

Exercise 1: The wheelbarrow and its contents have a mass of 50 kg and a center of mass at G. If the resultant moment produced by force F and the weight about point A (located at the center of the wheel) is to be zero, determine the required magnitude of force F.

Exeercise 2: Find the magnitudes of forces P and F to keep the block in Equilibrium.

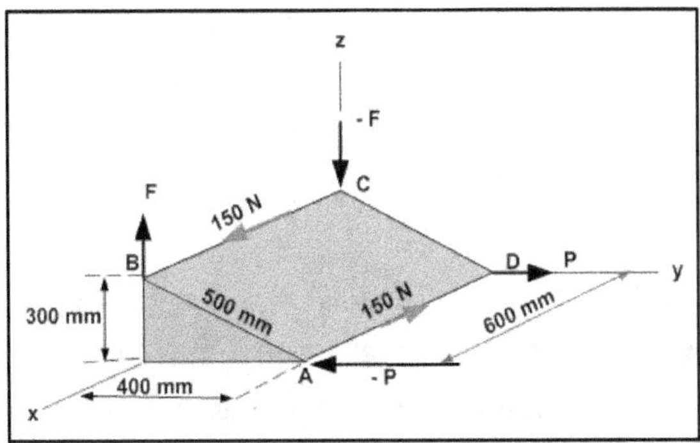

1.6. Friction: A Force that Opposes Sliding Relative Motion

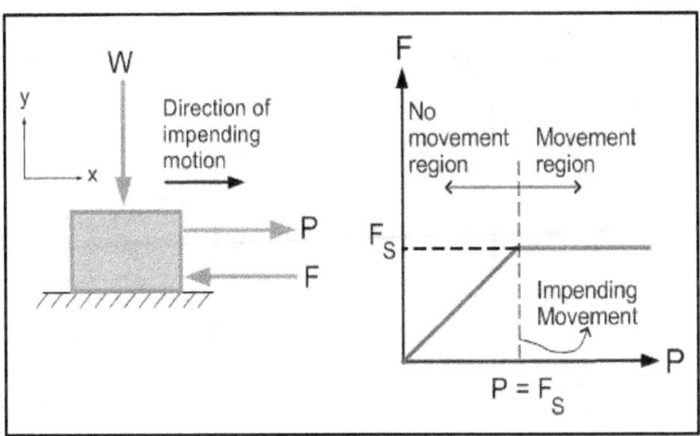

This module considers the phenomenon of friction in its simplest terms. Consider the case of a block resting on a surface and perform a thought experiment. Suppose a force P with very small magnitude is applied on the object as shown. It is common experience that for very small force P the object doesn't move. If we ponder as to why movement is not occurring, it becomes evident to us that some other force is also acting on the body and is cancelling the effects of P.

This force is what we call F, the friction force. If we very slightly increase the force P, the object doesn't move and it can be concluded that the force F also increased, giving us inference that **Friction is an automatic force**.

However, the friction force F does not increase unboundedly. As shown in the figure above, when the force P reaches a particular value $P=F_s$, friction force can no longer increase and movement of block impends[24]. Thus, we can imagine a region of "No movement" and a region of "movement" lying on either side of the "impending movement" line shown in Figure above. The force F_S is termed "limiting force of static friction". F_S depends on the materials of the bodies in contact and the total load W acting on the body.

Laws of Dry Friction

Guillaume Amontons and Charles-Augustin de Coulomb studied the dry friction phenomenon. Their experiments may be presented in terms of the following laws[25]:

1. The force of friction is directly proportional to the applied load.
2. The force of friction is independent of the apparent area of contact.
3. The limiting friction force generated is proportional to the normal force existing at contact surface.
4. Kinetic friction is independent of the sliding velocity.

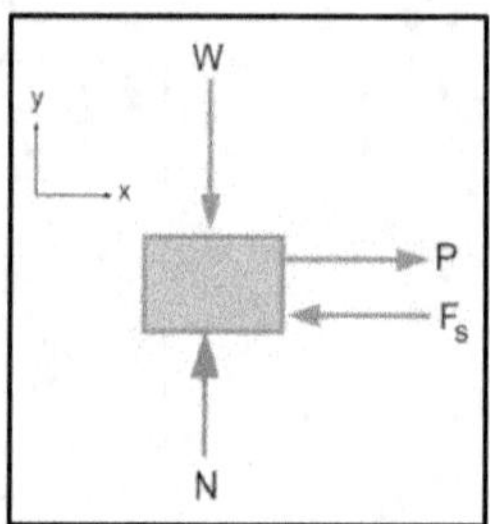

To illustrate the Law (3)(4), consider the free-body-diagram (FBD) of the block shown in diagram beside in which N is the normal force and F is the frictional force. Suppose that magnitude of P is just sufficient to cause sliding of the block.

Equilibrium equation along y-axis gives N = W. Law (3) gives, $F_s \propto N$ or $F_s = \mu_s N$ where μ_s = constant of proportionality and is termed static coefficient of friction. It may be noted very carefully that the equation $F_s = \mu_s N$ is applicable only when the motion is impending. In the region of no movement, F = P.

When once the sliding begins and the body is sliding, the friction force is slightly less than F_k and the corresponding coefficient of friction is called coefficient of kinetic friction, denoted by μ_k and we write $F_k = \mu_k N$.

---

[24]The word "impends" here is used to mean that the movement of the block did not begin but any small increase, however small it is, in the force will cause movement of the block.
[25]Guillaume Amontons also studied the phenomenon of friction much prior to Coulomb.

Angle of Repose

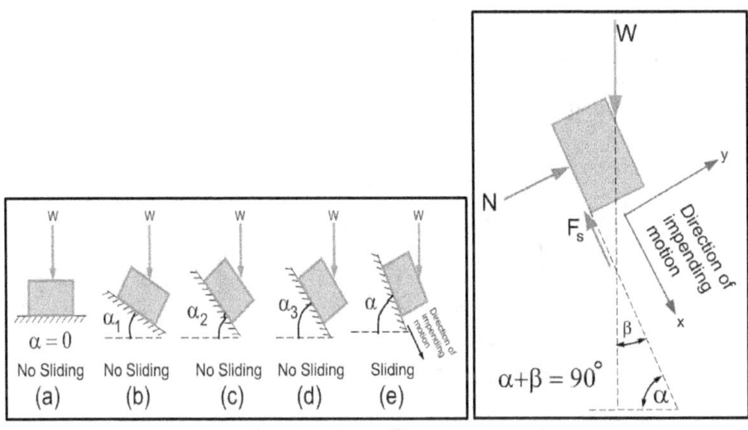

Consider a block of weight resting on level surface as shown in Figure (a) below. Let the coefficient of static friction between the material of the block and the material of the surface be μ_s. When the angle of inclination is $\alpha = 0$, obviously the block will not have any tendency to slide. When the angle of inclination is gradually increased to angles $\alpha_1 < \alpha_2 < \alpha_3$ ......, at some value of α, the block impends to slide. This angle is called angle of repose. To find the relation between the angle of repose and the static coefficient of friction μ_s, let us draw FBD when the angle of inclination is α and choose the x, y coordinate axes[26] as shown in the Figure beside and write the equations of equilibrium along these axes.

$$\sum F_x = 0 \Rightarrow W \cos\beta - F_s = 0 \Rightarrow F_s = W \sin\alpha$$

$$\sum F_y = 0 \Rightarrow W \sin\beta - N = 0 \Rightarrow N = W \cos\alpha$$

Dividing the first equation by the second equation, we obtain $\frac{F_s}{N} = \tan\alpha = \mu_s$ from the definition of static coefficient of friction $\therefore \alpha = \tan^{-1}(\mu_s)$. Remember the following points carefully before attempting to solve problems involving friction.

- If θ = inclination of the plane and α = angle of repose for the materials in contact,
  - $\theta < \alpha \Rightarrow$ block does not slide.
  - $\theta = \alpha \Rightarrow$ block impends to slide.
  - $\theta > \alpha \Rightarrow$ block is sliding.
- μ_s depends on the materials in contact.

---

[26]Why did we choose the coordinate axes in such an awkward way?

- **$\mu_s > 1$ for certain pairs of materials in contact.**

- μ_s decreases to μ_k when once the block starts sliding. μ_k is called kinetic friction.

- μ_s is independent of apparent area of contact.

Example: Coefficient of Friction

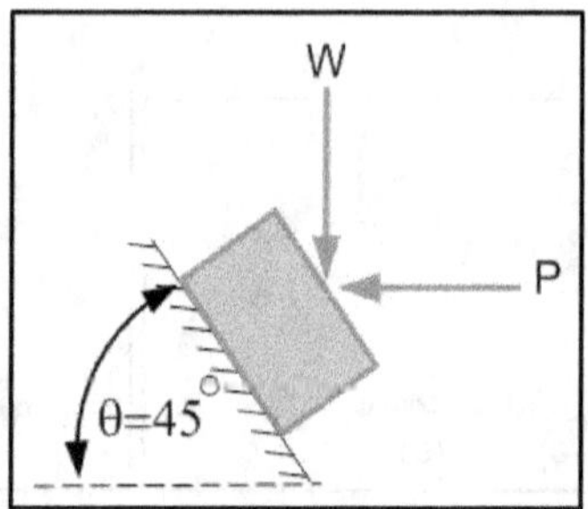

A block of 1000 N weight is put on an inclined plane making 45° with the horizontal. The coefficient of friction between the material of the block and the material of the plane is 0.58. Determine the force P needed to keep the block from moving down.

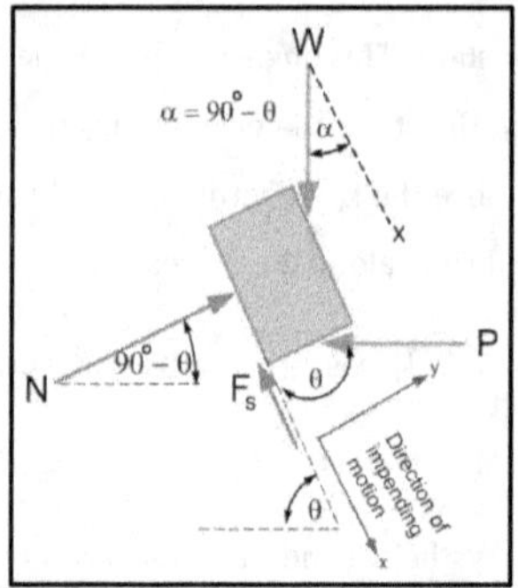

Solution: α = Angle of repose = $\tan^{-1}(0.58) \approx 30^{\circ}$. Since $\theta > \alpha$, the block will slide down if it is left on its own. We need to find the force P which is required to prevent the block from sliding. As a thought experiment, we will begin with P=0, for which case the block will slide down. We gradually increase the value of P till the block will just be on the verge of sliding down. At the instant the block is just on the verge of sliding down, we will draw the FBD of block in this situation as shown in Figure[27].

---

[27]It is simple but tricky to find the angles, and quadrant in which each force lies. Remember that the quadrant in which the force lies may easily be found by positioning the force such that its tail coincides with origin; the quadrant in which the arrowhead of the force lies gives the quadrant in which the force lies.

$$\sum F_x = 0 \Rightarrow -F_s - P\cos\theta + W\sin\theta = 0$$

$$\Rightarrow F_s = W\sin\theta - P\cos\theta$$

$$\sum F_y = 0 \Rightarrow N - W\cos\theta - P\sin\theta = 0$$

$$\Rightarrow N = W\cos\theta + P\sin\theta$$

From the friction equation, we have $F_s = \mu_s N \Rightarrow (W\sin\theta - P\cos\theta) = \mu_s(W\cos\theta + P\sin\theta)$.

$$\Rightarrow (\sin\theta - \mu_s\cos\theta)W = (\cos\theta + \mu_s\sin\theta)P$$

$$\Rightarrow P = \frac{(\sin\theta - \mu_s\cos\theta)}{(\cos\theta + \mu_s\sin\theta)}W = \frac{\tan\theta - \tan\alpha}{1 + \tan\theta\tan\alpha}W$$

Thus, $P = W\tan(\theta - \alpha) = 1000\tan 15\,^\circ = 268\ N$.

<u>Exercise</u>: For this problem, find the value of P required to initiate movement of the block up the plane.

1.7. Centroid of Plane Figures

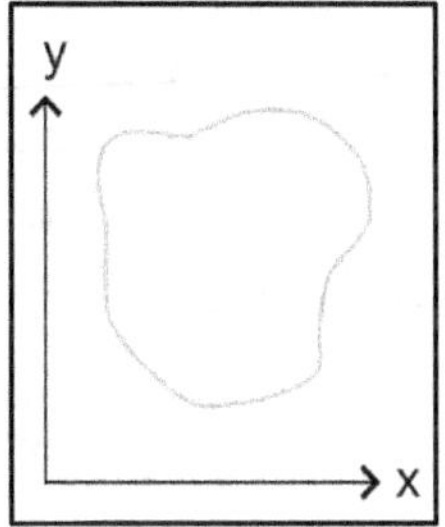

Centroid of a plane geometric figure is a point such that any straight line drawn through it will divide the figure into two equal areas.

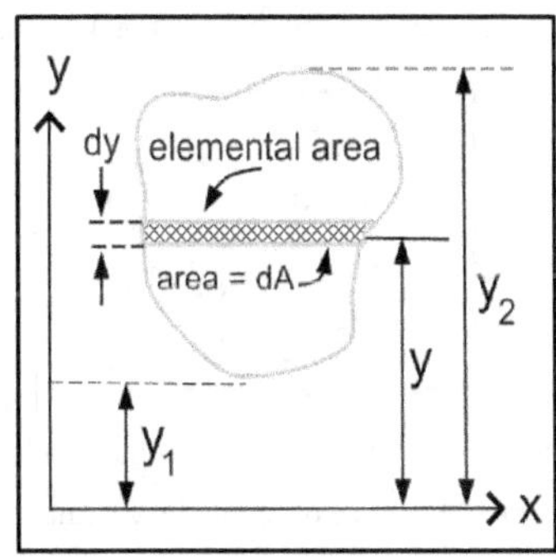

Consider the plane figure shown beside We may consider this as a thin sheet of negligible thickness cut into the shape. To find the **y-coordinate** of the shape, w.r.t. the x,y-axes shown, consider a thin strip of material **parallel to x-axis** as shown in Figure. The area of the strip is considered as "dA" and the strip is considered to be at a distance of "y" from x-axis. Then, the y-coordinate of the centroid of the plane figure, denoted by "y_c" is defined as:

$$y_c = \frac{\int_{y_1}^{y_2} y \cdot dA}{\int_{y_1}^{y_2} dA}.$$

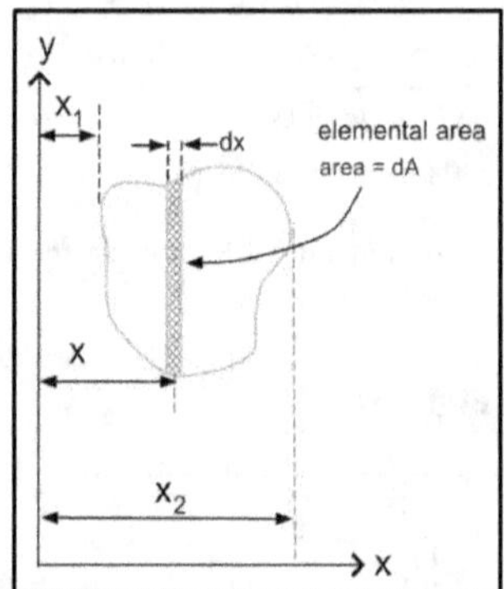

Similarly, to find the x-coordinate, we consider an elemental area parallel to y-axis and write the formula for x-coordinate of centroid, x_c , as:

$$x_c = \frac{\int_{x_1}^{x} x \cdot dA}{\int_{x_1}^{x} dA}.$$

The integration, sometimes, can be greatly simplified by choosing elemental area carefully.

Example: Centroid of Rectangle

Find the location of centroid of a rectangle with height of 4 cm and width of 6 cm.

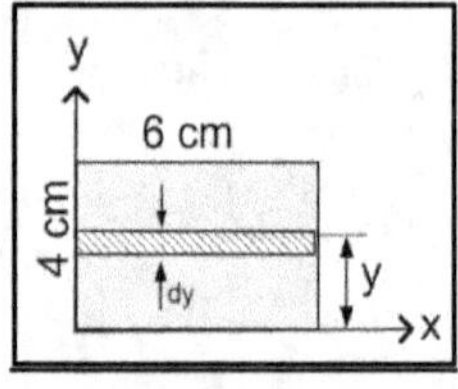

<u>Solution</u>: To find the y-coordinate of the centroid, choose an elemental area as shown. dA = 6(dy). Thus, y-coordinate of centroid may be obtained by writing:

$$y_c = \frac{\int_0^4 y \cdot dA}{\int_0^4 dA} = \frac{\int_0^4 y \cdot 6(dy)}{\int_0^4 6(dy)} = \frac{\left[\frac{y^2}{2}\right]_0^4}{[y]_0^4} = \frac{8}{4} = 2 \text{ cm}.$$

Similarly, the formula for y-coordinate of the centroid may be obtained by writing:

$$x_c = \frac{\int_0^6 x \cdot dA}{\int_0^6 dA} = \frac{\int_0^6 x \cdot 4(dx)}{\int_0^6 4(dx)} = \frac{\left[\frac{x^2}{2}\right]_0^6}{[x]_0^6} = \frac{18}{6} = 3 \text{ cm}.$$

Example: Centroid of Triangle

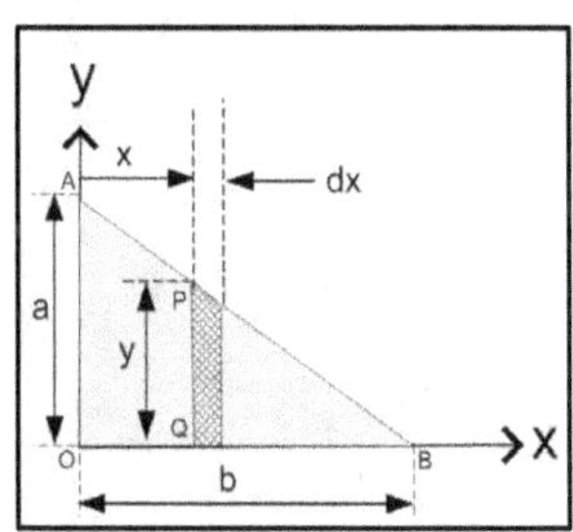

Locate the centroid of right angle triangle of base b and height a.

Solution: Choose x and y axes as shown in Figure . To find x_c , consider the elemental area as shown. With this choce of elemental area, dA = y(dx). To find "y" in terms of x, we notice that $\triangle BPQ \parallel\!\parallel \triangle BAO$. Therefore,

$$\frac{PQ}{OA} = \frac{BQ}{BO} \Rightarrow \frac{y}{a} = \frac{(b-x)}{b} \Rightarrow y = \frac{a}{b}(b - x).$$

Therefore, an expression for x_c is written as

$$x_c = \frac{\int_0^b x(dA)}{\int_0^b dA} = \frac{\int_0^b \frac{a}{b}(b-x)x(dx)}{\int_0^b \frac{a}{b}(b-x)(dx)} = \frac{\left[\frac{bx^2}{2} - \frac{x^3}{3}\right]_0^b}{\left[bx - \frac{x^2}{2}\right]_0^b} = \frac{\frac{b^3}{2} - \frac{b^3}{3}}{b^2 - \frac{b^2}{2}} = \frac{b}{3}$$

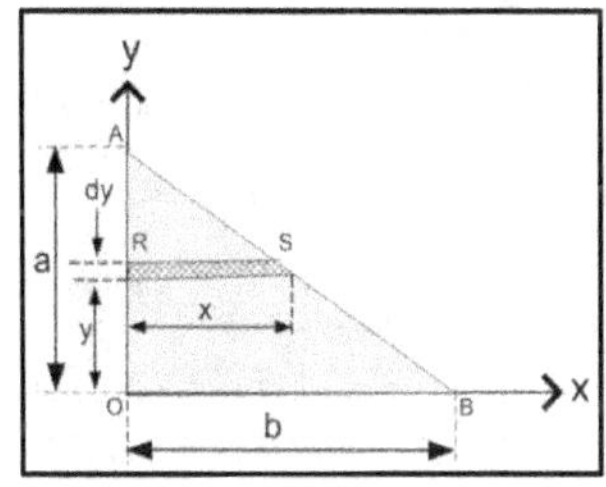

Similarly, to find y_c , we consider an elemental area as shown in figure. In the figure $dA = x(dy)$.

From the triangles $\triangle ARS \;\|\|\; \triangle AOB$ hence

$$\frac{x}{b} = \frac{(a-y)}{a} \Rightarrow x = \frac{b}{a}(a-y) \Rightarrow dA = \frac{b}{a}(a-y)dy.$$

Therefore y_c may be written as:

$$y_c = \frac{\int_0^a y(dA)}{\int_0^a dA} = \frac{\int_0^a \frac{b}{a}(a-y)y(dy)}{\int_0^a \frac{b}{a}(a-y)(dy)} = \frac{\left[\frac{ay^2}{2} - \frac{y^3}{2}\right]_0^a}{\left[ay - \frac{y^2}{2}\right]_0^a} = \frac{\frac{a^3}{2} - \frac{a^3}{3}}{a^2 - \frac{a^2}{2}} = \frac{a}{3}.$$

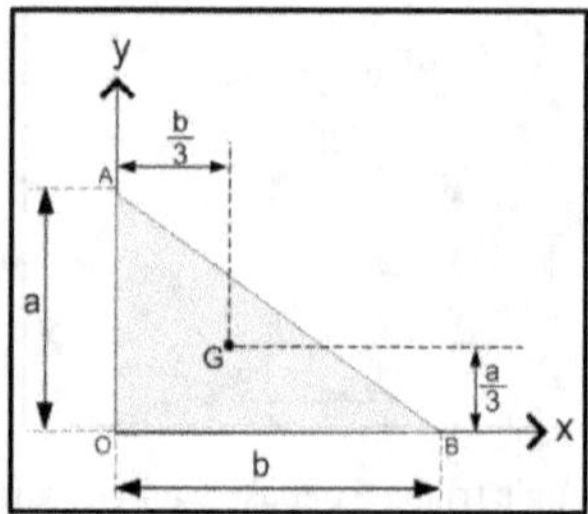

The centroid of a right angled triangle is located at G as shown in figure beside at $\frac{1}{3}rd$ the distance from each side.

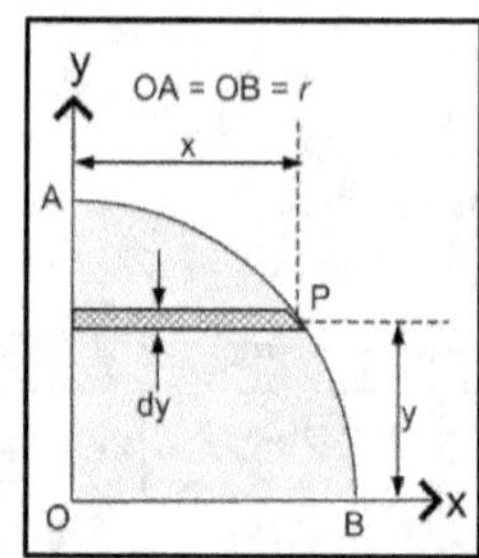

Example: Centroid of Quarter Circle

<u>Solution</u>: Since P is a point on circle, $x^2 + y^2 = r^2 \Rightarrow x = \sqrt{r^2 - y^2}$.

$$\therefore dA = x(dy) = \sqrt{r^2 - y^2}(dy) \Rightarrow y_c = \frac{\int_0^r y\sqrt{(r^2 - y^2)}(dy)}{\int_0^r \sqrt{r^2 - y^2}(dy)}$$

To evaluate numerator, let $r^2 - y^2 = t^2 \Rightarrow -2y(dy) = -t(dt)$.

When y=0, t=r. Similarly, when y=r, t=0. Therefore, numerator $= -\int_r^0 t^2(dt) = \frac{r^3}{3}$. To evaluate the denominator, let $y = r \sin \theta$ so that $dy = r \cos \theta \, (d\theta)$. When y=0. $\theta = 0$. When y=r, $\theta = \frac{\pi}{2}$. Therefore, denominator is evaluated as[28]:

$$\int_0^r \sqrt{r^2 - y^2}\,(dy) = \int_0^{\frac{\pi}{2}} (r \cos \theta)(r \cos \theta)(d\theta) = \frac{r^2}{2} \int_0^{\frac{\pi}{2}} (1 + \cos(2\theta))(d\theta) = \frac{\pi r^2}{4}.$$

$$\therefore y_c = \frac{r^3}{3} \times \frac{4}{\pi r^2} = \frac{4r}{3\pi}.$$

As an exercise, prove that $x_c = \frac{4r}{3\pi}$.

The following general method may be used as a check-list to ensure that no intermediate step is omitted:

1. Establish x and y axes and identify x_1, x_2, y_1, and y_2. Many times, it is very convenient to choose $x_1 = 0 = y_1$ by choosing axes to touch the geometric figure.
2. To find y-coordinate of centroid, consider the elemental area parallel to x-axis . Similarly, to find the x-coordinate, consider the elemental area parallel to y-axis.
3. Remember that any straight line passing through centroid divides the figure into two parts of equal area.
4. If a geometric figure has a line of symmetry, centroid must lie on that line of symmetry.

Centroid of Composite Areas

So far only simple shapes, such as a square, triangle, arc of a circle, which are called "primitives", are considered, coordinates of whose centroids were computed using integration. Often engineering objects comprise of combinations of primitives.

Once we know the locations of the centroids of the primitive objects, there is no need of performing integrations for finding the coordinates of composite object.

This aspect is presented in this module.

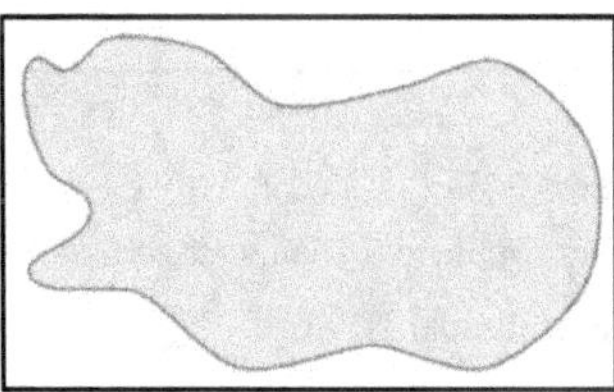

---

[28]Did you notice that this is the area of quarter circle?

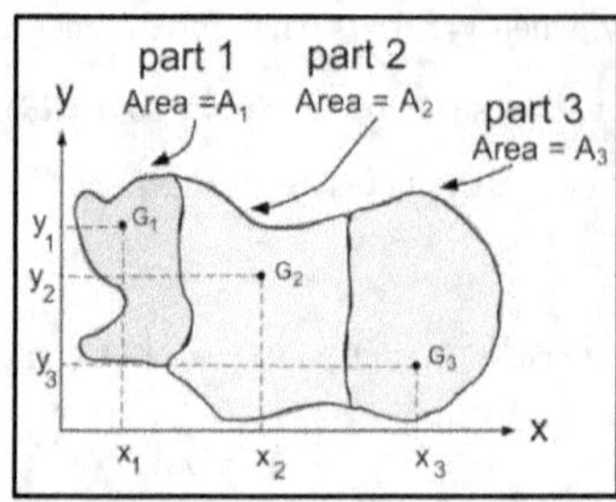

As an example, consider a hypothetical object shown in figure beside. This object can conceptually be broken into, say, three parts whose individual centroids are either mathematically computed or experimentally determined with respect to a coordinate system as shown in Figure. Let G_1, G_2, and G_3 be the centroids of the individual plane figures. Further, let x_1, x_2, x_3, y_1, y_2, and y_3 be the x and y coordinates of the points G_1, G_2, and G_3 in the coordinate system shown by x,y axes. If the areas of the segments of the plane figures of A_1, A_2, and A_3, the coordinates of the centroid of composite area may be found as:

$$x_c = \frac{A_1 x_1 + A_2 x_2 + A_3 x_3}{A_1 + A_2 + A_3}$$

$$y_c = \frac{A_1 y_1 + A_1 y_1 + A_3 y_3}{A_1 + A_2 + A_3}$$

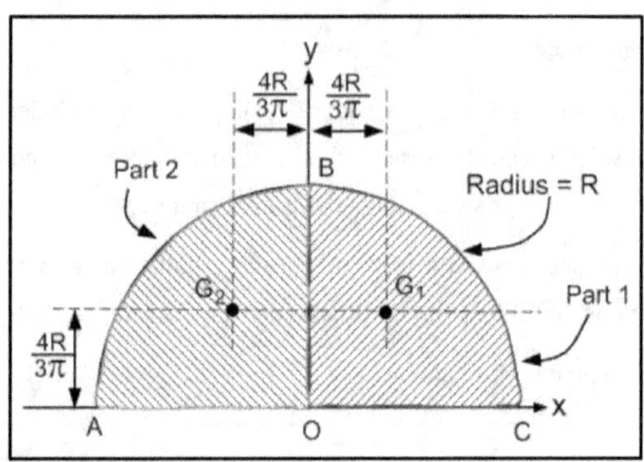

Example: Composite Areas

A weld ment is built from two quarter-circular sheets as shown in the figure. Find the location of the centroid of the semicircular sheet.

Solution: Such problems on composite objects are best solved using tabulation method. Carefully notice the tabular form and see how individual entries in the table are computed.

S.No.	Object	Area, A_i	x-coordinate	y-coordinate	$A_i x_i$	$A_i y_i$
1	OBC	$\dfrac{\pi R^2}{4}$	$\dfrac{4R}{3\pi}$	$\dfrac{4R}{3\pi}$	$\dfrac{R^3}{3}$	$\dfrac{R^3}{3}$
2	OAB	$\dfrac{\pi R^2}{4}$	$-\dfrac{4R}{3\pi}$	$\dfrac{4R}{3\pi}$	$-\dfrac{R^3}{3}$	$\dfrac{R^3}{3}$
Sum		$\dfrac{\pi R^2}{2}$ $(A_1 + A_2)$			0 $(A_1 x_1 + A_2 x_2)$	$\dfrac{2R^3}{3}$ $(A_1 y_1 + A_2 y_2)$

Therefore, coordinates of centroid may be written as[29]:

$$x_c = \frac{A_1 x_1 + A_2 x_2}{A_1 + A_2} = \frac{0}{\left(\dfrac{\pi R^2}{2}\right)} = 0$$

$$y_c = \frac{A_1 y_1 + A_2 y_2}{A_1 + A_2} = \frac{2R^3}{3} \times \frac{2}{\pi R^2} = \frac{4R}{3\pi}$$

Example: Composite Area with Hole

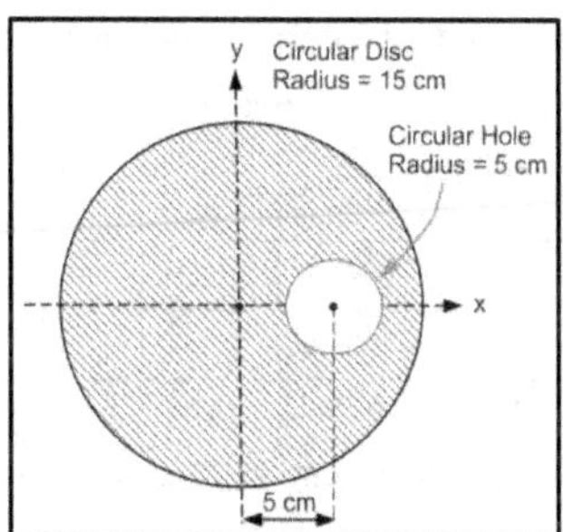

A sheet metal of thickness 0.50 mm thickness is cut into the form of a disc and a hole is drilled into it as shown in figure beside. Find the location of the centroid of the sheet.

Solution

S.No.	Object	A_i (cm²)	x_i	y_i	$A_i x_i$	$A_i y_i$
1	Disc	706.9	0	0	0	0
2	Hole	-78.5	5	0	-392.7	0
	Sum	628.3			-392.7	0

Therefore, the coordinates of the centroid of the disc with hole are obtained as

$$x_c = \frac{A_1 x_1 + A_2 x_2}{A_1 + A_2} = \frac{-392.7}{628.3} = -0.625 \text{ cm and}$$

---

[29]We need not compute x_c because the centroid lies on y-axis since y-axis is axis of symmetry. Obviously x_c=0.

$$x_c = \frac{A_1 y_1 + A_2 y_2}{A_1 + A_2} = \frac{0}{628.3} = 0 \text{ cm}$$

Example 2: Composite Area with Hole

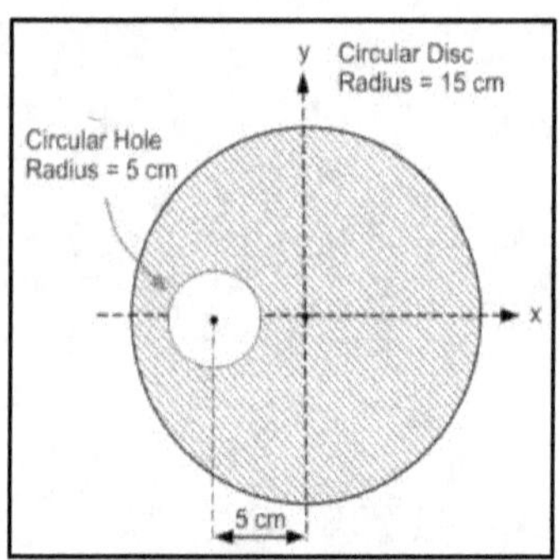

A sheet metal of thickness 0.5 mm thickness is cut into the form of a disc and a hole is drilled into it as shown in Figure beside. Find the location of centroid of the sheet.

S.No.	Object	A_i (cm²)	x_i	y_i	$A_i x_i$	$A_i y_i$
1	Disc	706.9	0	0	0	0
2	Hole	-78.5	-5	0	392.7	0
Sum		628.3			392.7	0

Therefore, the coordinates of the centroid of the disc with hole are obtained as

$$x_c = \frac{A_1 x_1 + A_2 x_2}{A_1 + A_2} = \frac{392.7}{628.3} = 0.625 \text{ cm and}$$

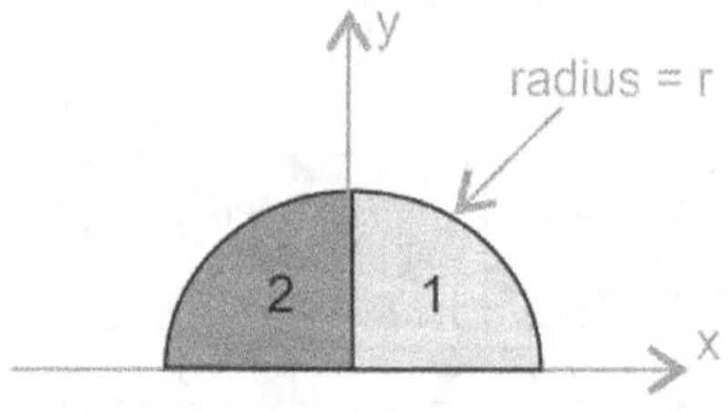

Fig.3

$$x_c = \frac{A_1 y_1 + A_2 y_2}{A_1 + A_2} = \frac{0}{628.3} = 0 \text{ cm}$$

Observe closely that the centroid shifted in a direction opposite to that of the hole.

Example: Semicircular Arc

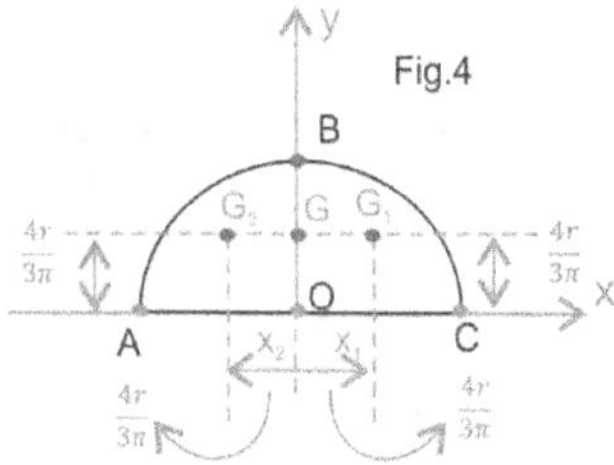

Divide the semicircle into two parts as shown in Fig.3 and establish x, y axes as shown, note that we have taken y-axis as axis of symmetry.

As seen in Fig.4, $x_1 = \frac{+4r}{3\pi} = -x_2 = y_1 = y_2$ Where x_1, y_1, x_2 and y_2 are coordinates of part 1 and part 2. Now, $x_c = \frac{A_1 x_1 + A_2 x_2}{A_1 + A_2} 0;\quad y_c = \frac{A_1 y_1 + A_2 y_2}{A_1 + A_2}$

So, coordinates of G are $(x_c, y_c) = \left(0, \frac{4r}{3\pi}\right)$

Through the solution presented earlier is correct, it is very convenient if the information is tabulated as shown in Table 1 below. Refer to Fig.2 for names of vertices etc.

Sl.No.	Object	Area A_i	x-Coordinate x_i	y-coordinate y_i	$A_i x_i$	$A_i y_i$
1	OAB	$\frac{\pi r^2}{4}$	$\frac{4r}{3\pi}$	$\frac{4r}{3\pi}$	$\frac{r^3}{3}$	$\frac{r^3}{3}$
2	OBC	$\frac{\pi r^2}{4}$	$-\frac{4r}{3\pi}$	$\frac{4r}{3\pi}$	$-\frac{r^3}{3}$	$\frac{r^3}{3}$
		$\frac{\pi r^2}{2}$			0	$\frac{2r^3}{3}$

$A_1 + A_2$

$A_2 y_2$

$A_1 x_1 + A_2 x_2$

$A_1 y_1 +$

$$\therefore x_c = 0 \text{ and } y_c = \frac{2r^3}{3} * \frac{2}{\pi r^3} = \frac{4r}{3\pi}$$

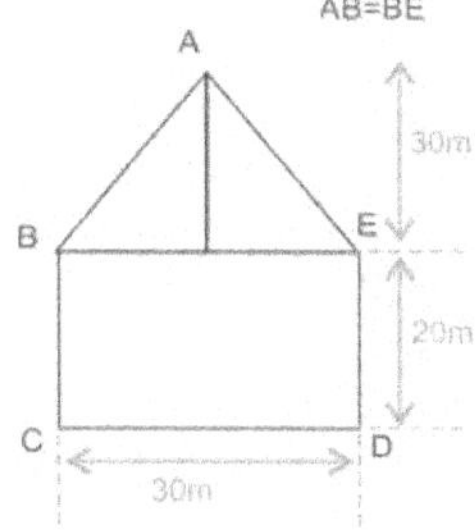

Example: Three Line Areas

Find centroid of composite object shown in Figure.

Solution: Divide the figure into 3 parts as shown in Figure 6 and establish X, Y axes as shown in Figure below.

Part 1: ABF

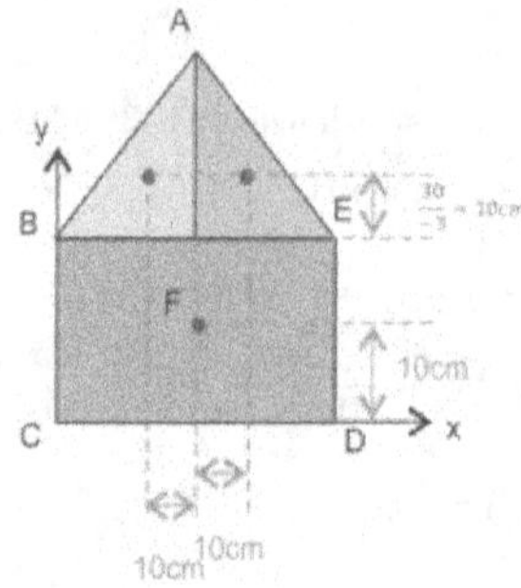

Part 2: AEF

Part 3: BCDE

Sl. No.	Object	Area	x_i	y_i	$A_i x_i$	$A_i y_i$
1	BCDE	600 cm²	15 cm	10 cm	9000	6000
2	ABF	225 cm²	10 cm	30 cm	2250	6750
3	AFE	225 cm²	20 cm	30 cm	4500	6750
		1050			15750	19500

$$x_c = 15 \ cm = \frac{15750}{1050}$$

$$y_c = 18.57 \ cm = \frac{19500}{1050}$$

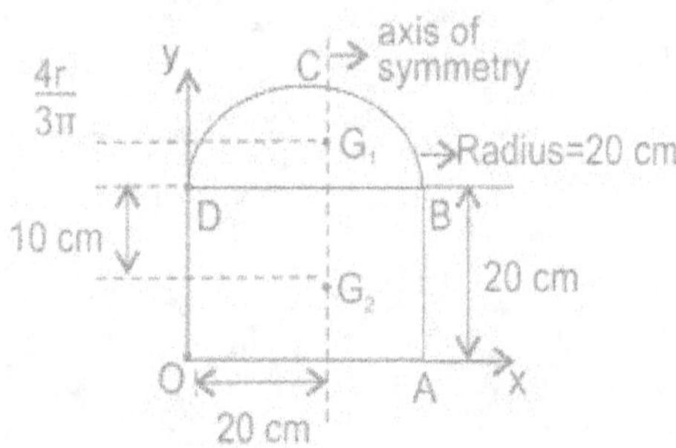

Example: Semicircle and Rectangle

Find the location of centroid of the shape shown in Figure beside.

Solution: There are two known shapes in Fig.3: Semicircle CBD and rectangle OABD. Their centroids and centroid of the composite figure OABCD, G, lie on the axis of symmetry. G, the centroid of BCD, is located at height of $\frac{4\pi}{3\pi} = \frac{80}{3\pi} = 8.48\ cm$ above BD. With this information, we can now build the Table as shown below:

Sl.No.	Object	Area A_i	x_i	y_i	$A_i x_i$	$A_i y_i$
1	BCD	628.3 cm²	20 cm	28.48 cm	12566 cm³	17894 cm³
2	OABD	800 cm²	20 cm	10 cm	16000 cm³	8000 cm³
		1428.3 cm²			28566 cm³	25894 cm³

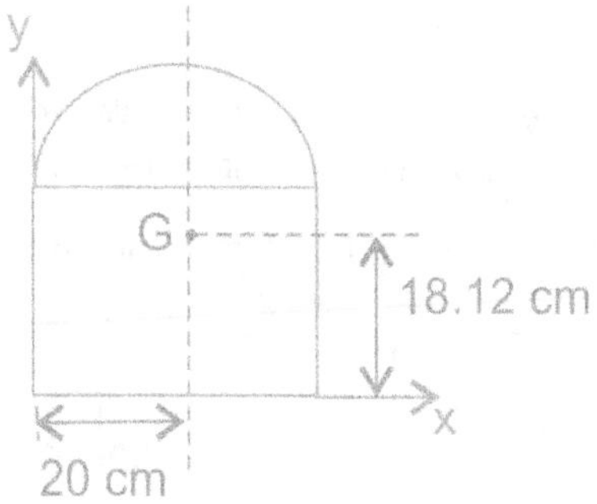

BCD Area $= \frac{\pi(20)^2}{2}$

OABD Area $= 40 * 20$

$\therefore\ x_c = \frac{28566}{1428.3} = 20cm$ [lying on Axis of Symmetry. Checked!]

$\&\ y_c = \frac{25894}{1428.3} = 18.12cm$

Fig.5

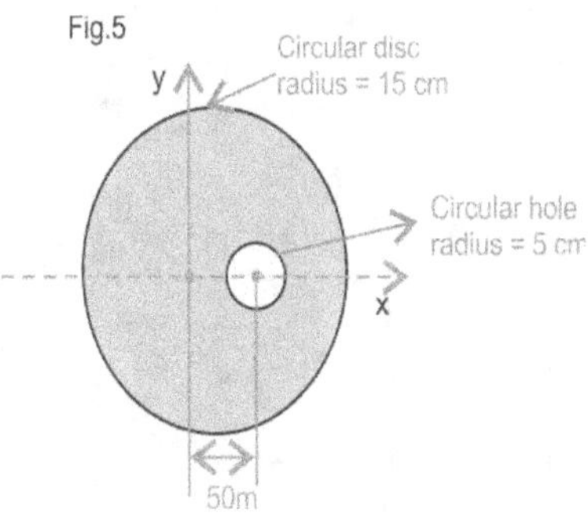

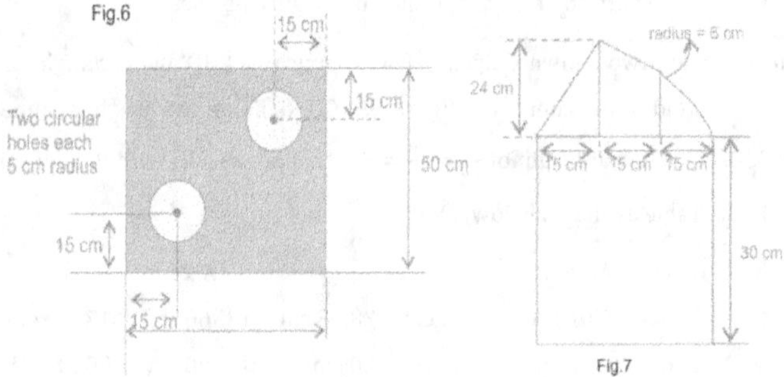

Exercise Problem

Find the location of centroid of following shaded areas:

1.8.   Centroid of Line Objects

Line objects do not have width or thickness. One particular type of line is straight line. The centroid of a straight line is located at the "middle" of that line.

Carefully notice the coordinates of centroids of straight lines in Figures 1(a), 1(b) and 1(c).

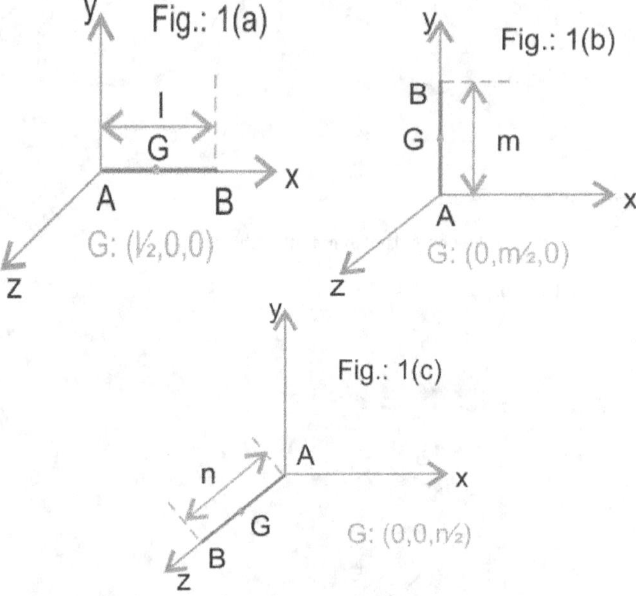

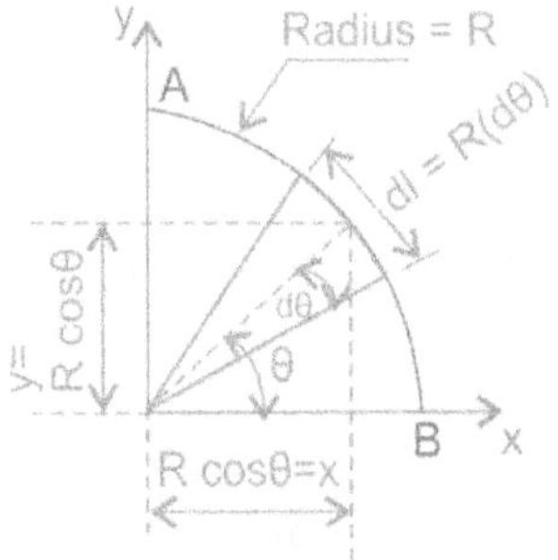

For the case of line segment joining points A and B with coordinates (x_1, y_1, z_1) and (x_2, y_2, z_2) the centroid is located at $\left(\frac{x_1+x_2}{2}, \frac{y_1+y_2}{2}, \frac{z_1+z_2}{2}\right)$. To determine the location of centroid of a curved line, small, infinitesimal segments are considered and each of which is assumed to be "straight". Example below explains the method.

Example: Circular Arc

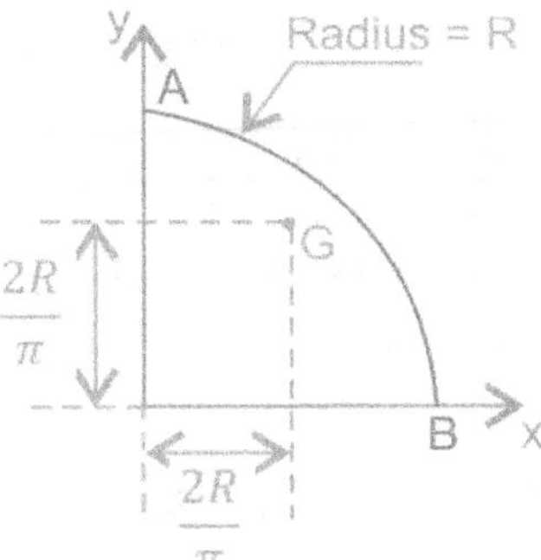

Determine the location of centroid of Quarter circle AB. Consider small element of length dl with centroid located at $(R \cos \theta, R \sin \theta)$ as shown.

Then, the length $dl = R(d\theta)$. We write the formula for coordinates of centroid of AB as:

$$\bar{x} = \frac{\int_0^{\frac{\pi}{2}} x \, dl}{\int_0^{\frac{\pi}{2}} dl} = \frac{\int_0^{\frac{\pi}{2}} (R \cos \theta)(R. \, d\theta)}{\int_0^{\frac{\pi}{2}} R \, d\theta} = \frac{R^2}{R.^{\pi}/_2} = \frac{2R}{\pi}$$

$$\bar{y} = \frac{\int_0^{\frac{\pi}{2}} dl}{\int_0^{\frac{\pi}{2}} dl} = \frac{\int_0^{\frac{\pi}{2}} (R \cos \theta)(R. \, d\theta)}{\int_0^{\frac{\pi}{2}} R \, d\theta} = \frac{R^2}{R.^{\pi}/_2} = \frac{2R}{\pi}$$

Centroid is shown as point G in Figure above.

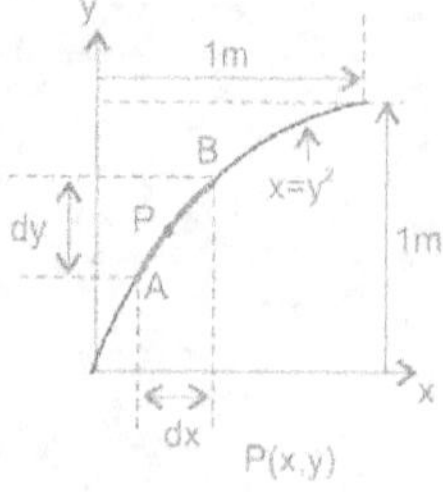

<u>Example</u>: Bent Rod

Locate the centroid of the rod bent into the shape of a parabolic arc.

<u>Solution</u>: Consider a small element AB with centroid at $P(x, y)$ and length dl. We can see that,

$$(dl)^2 = (dx)^2 + (dy)^2 \text{ and so } dl = \sqrt{(dx)^2 + (dy)^2} = \left\{\sqrt{1 + \left(\frac{dx}{dy}\right)^2}\right\} dy = \sqrt{(2y)^2}.\, dy$$

So, we write the formulae for coordinates of the centroid as:

$$\bar{x} = \frac{\int_0^1 x(dl)}{\int_0^1 dl} = \frac{\int_0^1 x\sqrt{1+4y^2}\,dy}{\int_0^1 \sqrt{1+4y^2}\,dy} = \frac{\int_0^1 y^2\sqrt{1+4y^2}\,dy}{\int_0^1 \sqrt{1+4y^2}\,dy} = \frac{0.61}{1.48} = 0.41m$$

$$\bar{y} = \frac{\int_0^1 y(dl)}{\int_0^1 dl} = \frac{\int_0^1 y\sqrt{1+4y^2}\,dy}{\int_0^1 \sqrt{1+4y^2}\,dy} = \frac{0.85}{1.48} = 0.57m$$

If a line has several segments of lengths $L_1, L_2, \dots\dots\dots, L_n$ and the centroids of the segments are located are $G_1(x_1, y_1, z_1)$, $G_2(x_2, y_2, z_2), \dots\dots\dots\dots, G_n(x_n, y_n, z_n)$ then, the centroid of the whole line is given as:

$$\bar{x} = \frac{L_1 x_1 + L_2 x_2 + \dots\dots + L_n x_n}{L_1 + L_2 + \dots\dots + L_n}, \bar{y} = \frac{L_1 y_1 + L_2 y_2 + \dots\dots + L_n y_n}{L_1 + L_2 + \dots\dots + L_n}, \bar{y} = \frac{L_1 z_1 + L_2 z_2 + \dots\dots + L_n z_n}{L_1 + L_2 + \dots\dots + L_n}$$

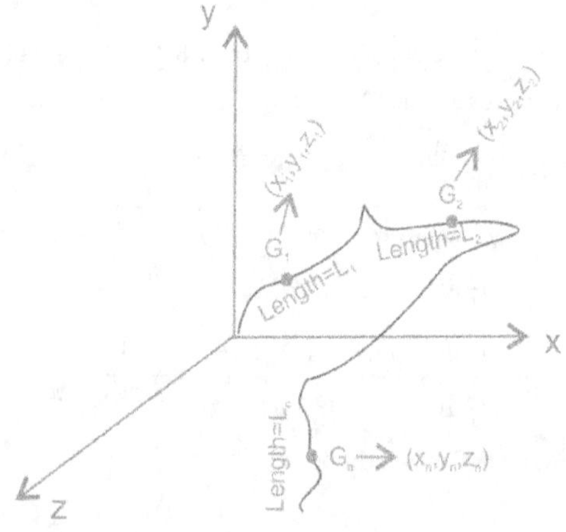

Example: Wire Bent into Two Planes

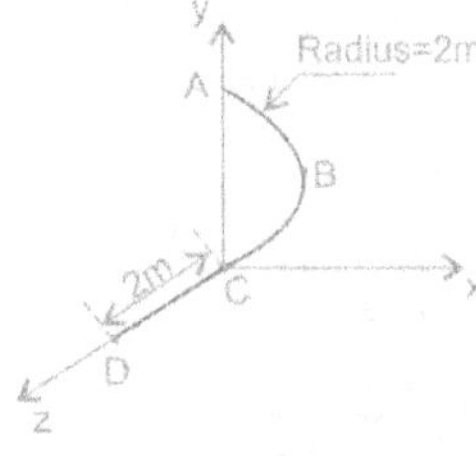

Locate the centroid of the wire bent as shown in Figure. The semi-circular portion of radius $R = 2m$ is lying in the xy-plane and the straight portion of length $2m$ is along the positive z-axis.

These are 3 segments:

- Quarter Circle AB
- Quarter circle BC
- Straight segment CD

The coordinates of centroids are:

$$G_1: \left(\frac{2R}{\pi}, R + \frac{2R}{\pi}, 0\right)$$

$$G_2: \left(\frac{2R}{\pi}, R - \frac{2R}{\pi}, 0\right)$$

$$G_3: (0, 0, 1)$$

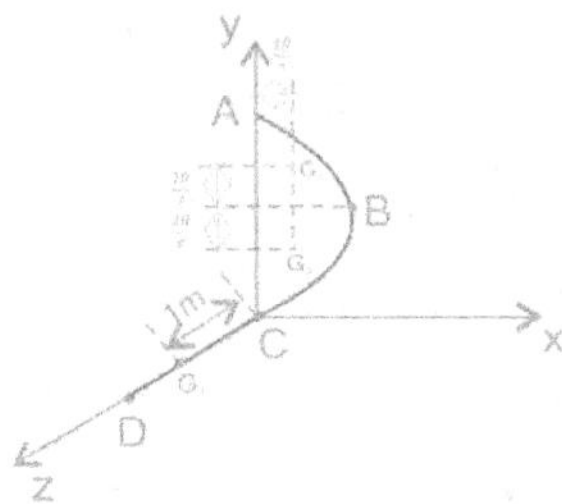

$$L_1 = Length\ of\ AB = \frac{\pi R}{2} = 3.14m = L_2$$

$$L_3 = 1m$$

$$\frac{2R}{\pi} = 1.27m$$

$$R + \frac{2R}{\pi} = 3.27m$$

$$R - \frac{2R}{\pi} = 0.73m$$

Sl.No.	Segment	Length (L_i, m)	$x_i(m)$	$y_i(m)$	$z_i(m)$	$L_i x_i$ (m^2)	$L_i y_i$ (m^2)	$L_i z_i$ (m^2)
1	AB	3.14	1.27	3.27	0	4	10.27	0
2	BC	3.14	1.27	0.73	0	4	2.30	0
3	CD	2	0	0	1	0	0	2
		8.28				8	12.57	2

$$\therefore \bar{x} = \frac{8}{8.28} = 0.97m; \ \bar{y} = \frac{12.57}{8.28} = 1.52m; \ \bar{z} = \frac{2}{8.28} = 0.241m$$

1.9. Centroid of Volume

The coordinates of centroid of a given solid are given as:

$$\bar{x} = \frac{\int_v xdv}{\int_v dv}, \bar{y} = \frac{\int_v ydv}{\int_v dv}, \bar{z} = \frac{\int_v zdv}{\int_v dv}$$

Where

$\bar{x}, \bar{y}, \bar{z}$ = coordinates of centroid

dv = elemental volume considered

x, y, z = coordinates of centre of element volume

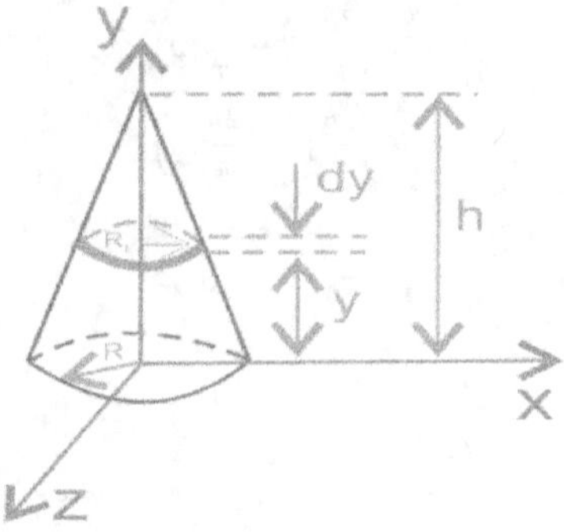

Example: Right Circular Cone

Determine the location of centroid of a right circular cone of base radius R and height h.

From geometry, we see $\frac{R_1}{R} = \frac{h-y}{h} \Rightarrow R_1 \frac{R}{h}(h-y)$

$$\therefore dv = \pi R_1^2 (dy) = \frac{\pi R^2}{h^2}(h-y)^2 dy$$

$$\therefore \bar{y} = \frac{\int y(dv)}{\int dv} = \frac{\int_0^h y\frac{\pi R^2}{h^2}(h-y)^2 dy}{\int_0^h \frac{\pi R^2}{h^2}(h-y)^2 dy} \longleftarrow \text{numerator} \\ \longleftarrow \text{denominator}$$

Numerator

$$\frac{\pi R^2}{h^2}\int_0^h y(h^2+y^2-2hy)dy=\left[\frac{h^2y^2}{2}+\frac{y^4}{4}-\frac{2hy^3}{3}\right]_0^h\frac{\pi R^2}{h^2}=\frac{\pi R^2}{h^2}\cdot h^4\left[\frac{1}{2}+\frac{1}{4}-\frac{2}{3}\right]=\frac{\pi R^2 h^2}{12}$$

Denominator:- $\frac{\pi R^2}{h^2}\int_0^h(h^2+y^2-2hy)=\frac{\pi R^2}{h^2}\cdot h^3\left[1+\frac{1}{3}-1\right]=\frac{1}{3}\pi R^2 h$

$$\therefore \bar{y}=\frac{\dfrac{\pi R^2 h^2}{12}}{\dfrac{\pi R^2 h}{3}}=\frac{h}{4}$$

Can you determine the values of $\bar{x}$ and $\bar{z}$?

Example: Centroid of Hemisphere

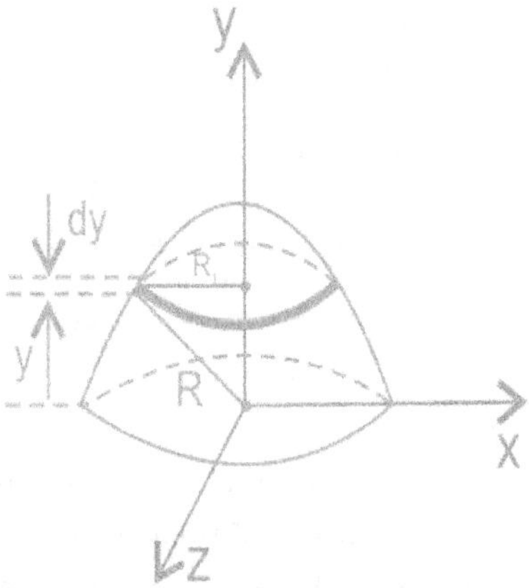

Consider elemental volume shown above. The elemental volume has center with coordinates $(0,y,0)$.

The radius of the disc is R_1 which can be found in terms of y and R from the equation $R_1^2+y^2=R^2$. So, $dv=$ elemental volume $=\pi(R^2-y^2)(dy)$. Therefore, we write

$$\bar{y}=\frac{\int y(dv)}{\int dv}=\frac{\int_0^R \pi y(R^2-y^2)(dy)}{\int_0^R \pi(R^2-y^2)dy}$$

Numerator:- $\int_0^R \pi y(R^2-y^2)dy=\frac{\pi R^4}{4}$

Denominator:- $\int_0^R \pi(R^2-y^2)dy=\frac{2}{3}\pi R^3$

$$\therefore \bar{x}=\frac{\pi R^4/4}{2\pi R^3/3}=\frac{3}{8}R$$

Can you find the values of $\bar{x}$ and $\bar{z}$?

Composite Bodies

As a natural extension, we see the following formulae to compute the coordinates of centroid of composite body made out of "primitive" shape.

$$\bar{x} = \frac{v_1 x_1 + \cdots \ldots + v_n x_n}{v_1 + \cdots \ldots + v_n}; \ \bar{y} = \frac{v_1 y_1 + \cdots \ldots + v_n y_n}{v_1 + \cdots \ldots + v_n}; \ \bar{z} = \frac{v_1 z_1 + \cdots \ldots + v_n z_n}{v_1 + \cdots \ldots + v_n}$$

Where $\bar{x}, \bar{y}, \bar{z}$ = coordinates of centroid of composite body

$v_1, \ldots \ldots, v_n$ = Volumes of individual "primitive bodies"

$x_1, \ldots \ldots, v_n$ = x − coordinates of centroids of parts

$y_1, y_2, \ldots \ldots, y_n$ = y − coordinates of centroids of parts

$z_1, z_2, \ldots \ldots, z_n$ = z − coordinates of centroids of parts

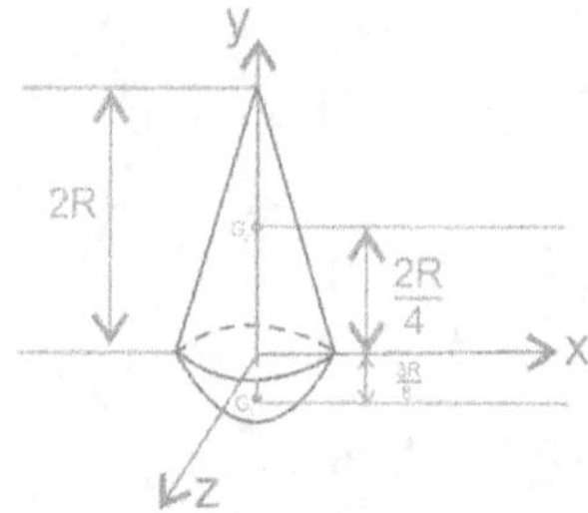

Example: Composite Solid

Determine the location of centroid of the solid shown. The radius of hemisphere is R and height of cone is 2R. Solution: The solid consists of a hemisphere and a cone choose xyz − axes in such a way that the xz − plane lies at the flat face of the hemisphere and origin coincides with the center of the flat face of the cone. Then, we know that G_1, the centroid of hemisphere lies at a distance of $\frac{3R}{8}$ on the negative y − axis. Also, G_2, the centroid of cone lies at a distance of $\frac{2R}{4}$ on the positive y − axis. Since, G_1 and G_2 lie on y − axis, $x_1 = z_1 = x_2 = z_2 = 0$. So, we can write the table for calculating centroid as:

$$\therefore \bar{y} = \frac{\frac{1}{12}\pi R^4}{\frac{4}{3}\pi R^3} = \frac{R}{16}; \ \bar{x} = \bar{z} = 0$$

Sl.No.	Object	Volume v_i	y_i	$v_i y_i$
1	Hemisphere	$\frac{2}{3}\pi R^3$	$\frac{-3R}{8}$	$\frac{-\pi R^4}{4}$
2	Cone	$\frac{2}{3}\pi R^3$	$\frac{R}{2}$	$\frac{\pi R^4}{3}$
		$\frac{4}{3}\pi R^3$		$\frac{1}{12}\pi R^4$

Example: Composite Solid

A composite solid is built from a cube and a square pyramid as shown. Determine the location of centroid of this solid.

Solution: With the axes setup as shown, we see that

$x_1 = z_1 = x_2 = z_2 = 0$. Further, $y_1 = 3m$ and $y_2 = 1m$

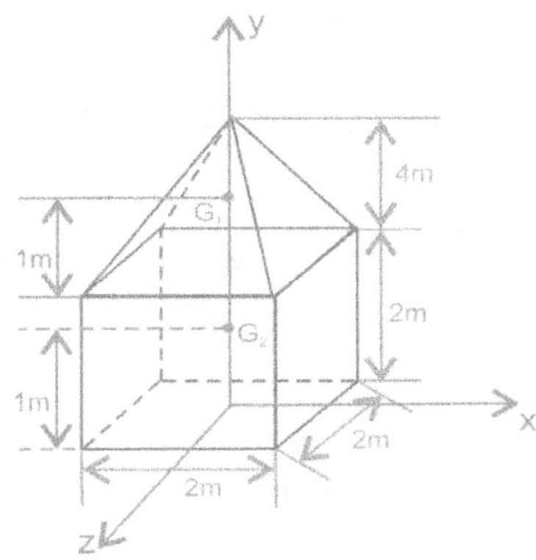

Sl. No.	Object	Volume $v_i(m^3)$	$y_i(m)$	$v_iy_i(m_4)$
1	Pyramid	$\dfrac{16}{3}$	3	16
2	Cube	8	1	8
		$\dfrac{40}{3}$		24

$\therefore \bar{y} = 24 \times \dfrac{3}{40} = 1.8m$

And $\bar{x} = \bar{z} = 0$

1.10. Centre of Gravity

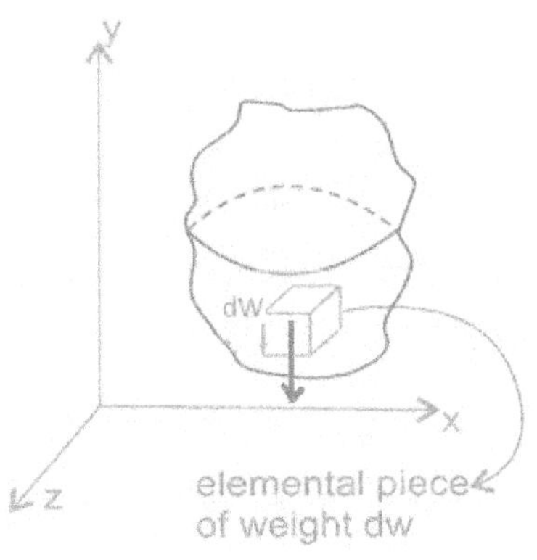

Center of Gravity (CG) is a point which locates the weight of a system of particles or bodies. The resultant of all gravity forces acting on all the particles comprising the system passthrough the CG. And hence, the sum of movements due to the individual particle weights about CG is equal to zero.

Consider a general body shown in figure 1. Within this body, consider an elemental particle of weight "dw" located at coordinates $x, y, and\ z$. With this setup, the x, y, z coordinates of CG may be given as:

$$x_c = \frac{\int x(dw)}{\int dw}, y_c = \frac{\int y(dw)}{\int dw}, z_c = \frac{\int z(dw)}{\int dw}$$

NOTE: The integration needs to be carried to "sweep" all the parts of the body. In a very general case this involves triple integral. However by carefully choosing cartesion / polar/spherical coordinate system and/or the choice of element carefully, the integration may be greatly simplified.

IF ALL THE REGIONS OF THE OBJECT HAVE EQUAL DENSITY AND if the object has an axis/plane of symmetry, then CG will lie on that line/plane.

Example: Centre of Gravity of a Rectangular Parallelepiped

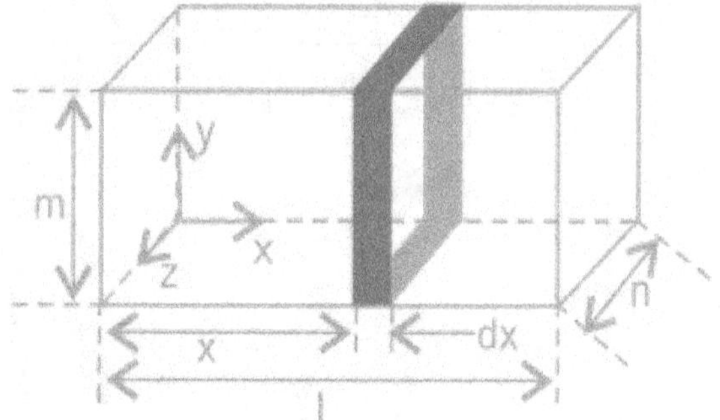

Consider an elemental "slice" perpendicular to $x -$ axis as shown in Figure 2. If the density of material of this object is ρ, then the weight of this slice is:

$dw = \rho(mn)(dx)g$ and the entire perallelopiped can be swept by considering slices at x distance for all values of x between zero and l. Thus, we write:

$$x_c = \frac{\int_0^l x.\rho(mn)(dx)g}{\int_0^l \rho(mn)(dx)g} = \frac{\int_0^l x(dx)g}{\int_0^l dx} = \frac{l}{2}$$

To convince yourself that you understood thoroughly, compute y_c & z_c.

Example: Centre of Gravity of a Right Circular Cone

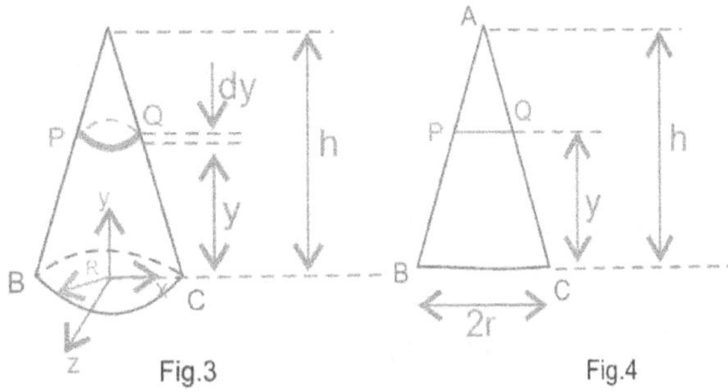

Fig.3   Fig.4

Solution:- The centre of gravity lies on y axis. Why?

Consider a circular slice at a height y and thickness dy. This slice will have a diameter of PQ as shown in Figure 4. Now, $\triangle APQ |||\triangle ABC$. Therefore, $\frac{PQ}{2r} = \frac{h-y}{h} \Rightarrow PQ = \frac{2r}{h}(h - y)$. Therefore, weight of the slice, assuming density of ρ is given by,

$$dw = \frac{\pi}{4}(PQ)^2(dy); \rho g = \frac{\pi\, \pi r^2}{4\, h^2}(h - y)^2 . \rho(dy)g$$

To cover the entire cone, we need to "sweep" the values of y from zero to h. Thus, y −coordinates of CG may be evaluated as:

$$y_c = \frac{\int_0^h y \frac{\pi r^2}{h^2}(h-y)^2 \rho(dy)g}{\int_0^h \frac{\pi r^2}{h^2}(h-y)^2 \rho(dy)g} = \frac{\int_0^h y(h-y)^2 dy}{\int_0^h (h-y)^2 dy} = \frac{h^4}{12} * \frac{3}{h^3} = \frac{h}{4}$$

Therefore, CG of cone lies at a distance of $\frac{h}{4}$ from the base on the axis.

Example: Centre of Gravity of a Hemisphere

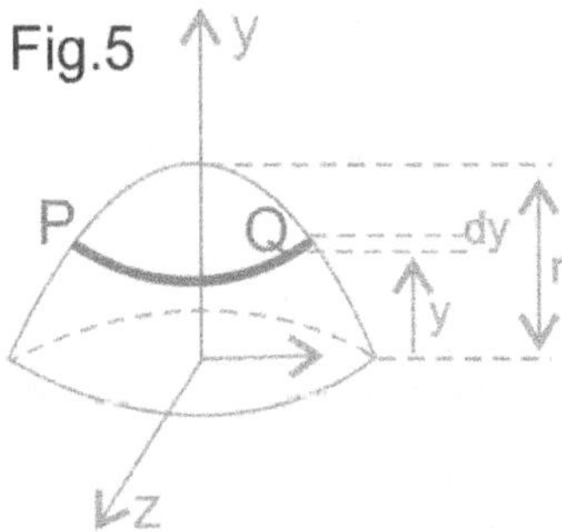

Fig.5

Let ρ be the density of the material. Consider a circular slice as shown in Figure 5. The radius of the circular slice is r_1 as shown in Figure 6 from which we can write,

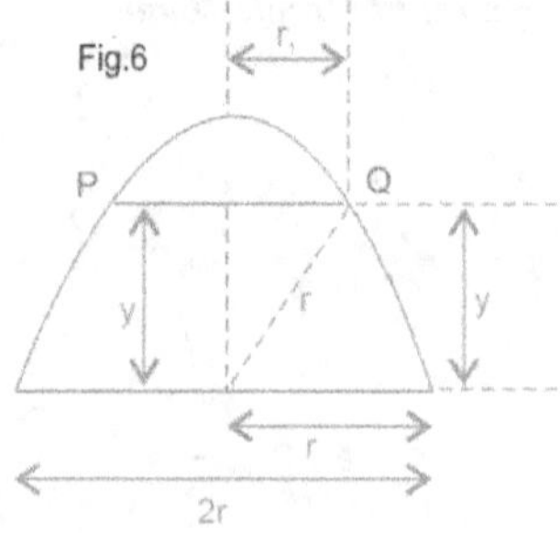

$$y^2 + r_1^2 = r^2 \ \ or \ \ r_1 = \sqrt{r^2 - y^2}.$$

So, the weight of the slice is given by,

$$dW = \pi r_1^2 (dy)\rho g = \pi \rho (r^2 - y^2)(dy)g$$

$\therefore y -$ coordinates CG may be calculated as:

$$y_c = \frac{\int_0^r y \pi \rho (r^2 - y^2)(dy)g}{\int_0^r \pi \rho (r^2 - y^2)(dy)g} = \frac{\int_0^r y(r^2 - y^2)dy}{\int_0^r (r^2 - y^2)dy} = \frac{r^4}{4} \times \frac{3}{2r^3} \frac{3}{8} r$$

[What about x_c and z_c?]

Therefore CG of hemi-sphere is located at a height of $\frac{3}{8}r$ from the flat face of the hemisphere.

1.11. Theorems of Pappus and Guldinus

Two Theorems to computer axisymmetric volumes and surface areas are attributed to: Pappus of Alexandria, 3 BC Greek Philosopher and Habakkuk Guldin, 1577 – 1643 AD.

Their first theorem allows us to compute the surface area of revolution of a line object and the second theorem allows us to compute the volume of solids obtained by revolving planes.

First Theorem of Pappus

The are of a surface of revolution equals the product of the length of the generating curve (L) and the distance traveled by the centroid of the curve while the area is being generated (D).

$$A_s = LD \qquad (1)$$

1.12. General Method for Finding Area of Revolution

1.   Locate / Identify the Generating curve.
2.   Find the length of the curve and locate centroid.
3.   Identify the axis of revolution.
4.   Find the angle of revolution and/or distance traveled by centroid.

5. Use equation (1)

Note that equation (1) applies both to plane areas and also areas of curved surfaces. It takes a bit of practice and visualization to locate the generating curve and axis of revolution.

Fig.1

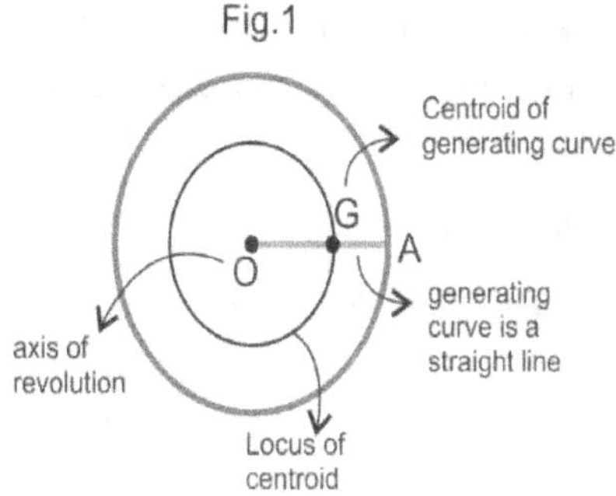

Example: Area of a Circle

The circle of radius "r" is generated by straight line OA as the line OA revolves about an axis perpendicular to the plane of the circle and passing the center of the circle. The centroid of generating line lies at G, midway between O and A. As the line OA rotates and "sweeps" the circular area, centroid travels along a circle of radius $\frac{r}{2}$. So, we note that

- OA is generating curve

- Length of generating curve, L=r

- Axis of revolution is ⊥ to plane of circle passing through O

- As OA sweeps circular area, G goes around a circular path, D = circumference of circular path of $G = 2\pi \left(\frac{r}{2}\right)$

- $\therefore A = L * D = r * 2\pi \left(\frac{r}{2}\right) = \pi r^2$

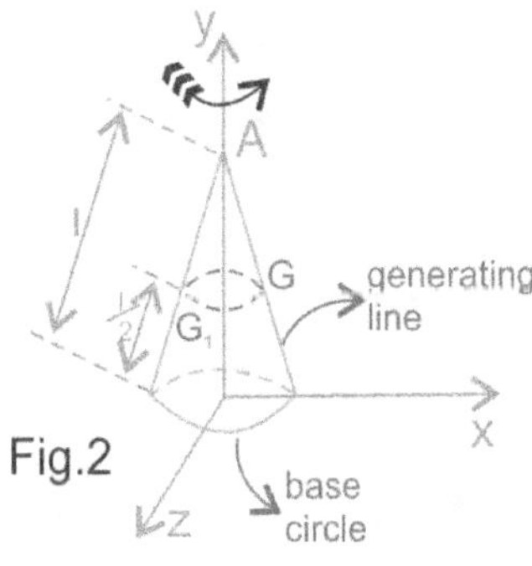

Fig.2

Example: Area of Curved Face of Cone

Figure 2 shows the cone in $xyz-$ axes set up. We see that axis of revolution is $y-$axis, generating curve is any one slant line on curved face of the cone joining the apex and a point on base circle. When the generating line is rotated about $y-$axis, the centroid of generating line, G, goes along a circular path as shown in Figure 2, the radius of which is shown as the line CG in Figure 3 from which we note:

$$\Delta ACG |||\Delta OAB \qquad \therefore \frac{r_1}{r} = \frac{AG}{AB} = \frac{1}{2} \qquad \therefore r_1 = \frac{r}{2}$$

L = Length of generating curve = $L = \sqrt{r^2 + h^2}$

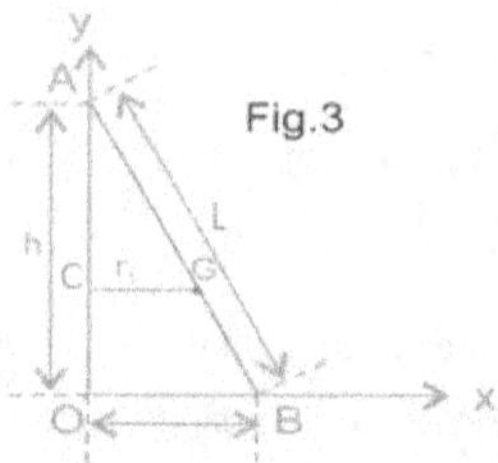

D = {Distance travel by centroid while generating conical surface} = Circumference of circle GG_1G $(in\ Figure\ 1) = 2\pi r_1, = 2\pi \frac{r}{2} = \pi r$

$\therefore$ non-flat area of cone = $L * D = \pi r \sqrt{r^2 + h^2}$

Second Theorem of Pappus

The volume of a body of revolution equal the generating area times the distance traveled by the centroid of the area while the body is being generated:

$$V = A * D$$

A = Generating area, D = Distance travelled by centroid, V = Volume of solid

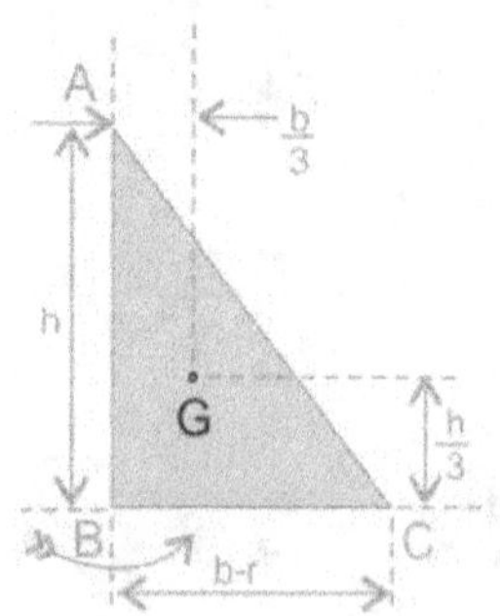

Example: Volume of Right Circular Cone

Right circular cone is generated by right angle triangle ABC as it is revolved about the side AB through an angle of 2π radius. While the cone is being generated, the centroid of the triangle, G, traverses a circular path with a radius of $\frac{b}{3}$.

- If the base circle radius is r, then $b = r$.

- Generating area $= \frac{1}{2}hb = \frac{1}{2}hr = A$

- Distance travelled by $G = 2\pi\left(\frac{b}{3}\right) = 2\pi\left(\frac{r}{3}\right) = D$

- Volume of cone $= A * D = \frac{1}{2}hr * 2\pi\left(\frac{r}{3}\right) = \frac{1}{3}\pi r^2 h$

We note that the procedure for computing volume of axisymmetric solid is similar to that of axisymmetric area except that generating area is used instead of generating curve.

1.13. Moment of Inertia

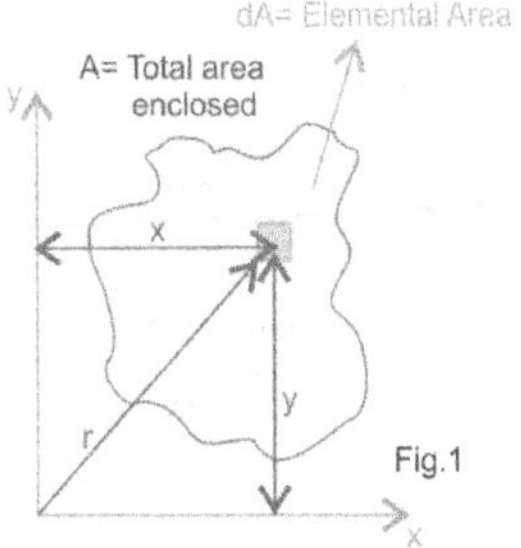

Moment of inertia, also called second moment of area is defined as integral of the product of the area and the square of its moment arm from the reference axis. With respect to Fig. 1, we define moment of Inertia of the area enclosed in the curve with respect to x-axis is given by:

$$I_x = \int_A y^2\,(dA) \quad (1)$$

Note that x-axis is the reference axis and "y" is the distance of elemental area measured from x-axis. Similarly, moment of Inertia of the area with respect to y-axis is given by:

$$I_y = \int_A x^2\,(dA) \quad (2)$$

In equations (1) and (2), integration needs to be performed over the entire area i.e., infinitely many infinitesimally small areas "dA" need to be considered and the term $K^2(dA)$ where K is the distance from reference axis (K = y in equation 1 and K = x in equation 2) need to be summed up: A term closely related to moment of inertia is radius of gyration. Equations (3) and (4) may be borne in mind:

$$I_x = K_x^2 A \Rightarrow K_x = \sqrt{\frac{I_x}{A}} \qquad (3)$$

$$I_y = K_y^2 A \Rightarrow K_y = \sqrt{\frac{I_y}{A}} \qquad (4)$$

Refer to fig. 1 and note the z-axis is perpendicular to both x and y axes (that is perpendicular to the plane of curve) passing through the origin and pointing towards the viewer. We can also see that "r" is the distance of elemental area from z-axis Moment of inertia w.r.t. z-axis may be written as:

$$I_z = \int_A r^2 (dA) = \int_A (x^2 + y^2) dA = I_x + I_y = J \ (5)$$

I_z is termed "polar" moment of inertia and is commonly denoted by symbol J so we write:

$J = \int_A r^2 dA = I_x + I_y$. We also define polar radius of gyration, K_z as:

$$K_z = \sqrt{\frac{J}{A}} = \sqrt{\frac{I_z}{A}} = \sqrt{\frac{I_x + I_y}{A}} \Rightarrow J = I_z = K_z^2 A \ (6)$$

Important

$$I_x > 0, I_y > 0, I_z > 0, J > 0$$

1. Values of I_x, I_y and I_z depend on the shape of the closed curve and the choice of location of origin and the $x, y, z -$ axes.

2. Of particular interest to us are the values of Moments of Inertia w.r.t. axes passing through centroid od area. These values are called centroidal Moments denoted by $I_{\bar{x}}, I_{\bar{y}}, I_{\bar{z}}$

Example: Moment of Inertia of Rectangle

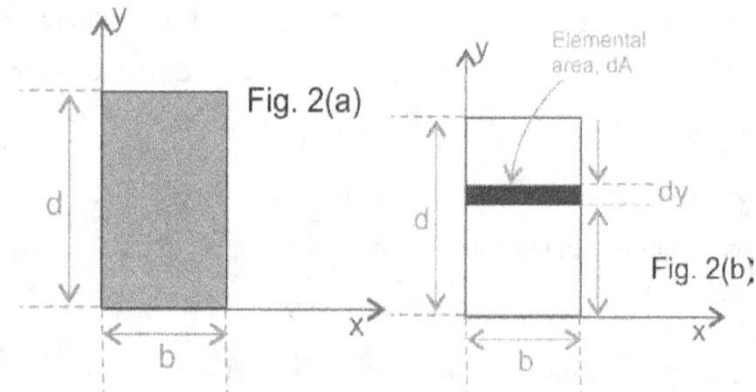

To find I_x:- Consider elemental area as shown in Figure 2(b)

$$dA = b(dy)$$

$$\therefore I_x = \int_0^d y^2 . b . dy = \frac{bd^3}{3}$$

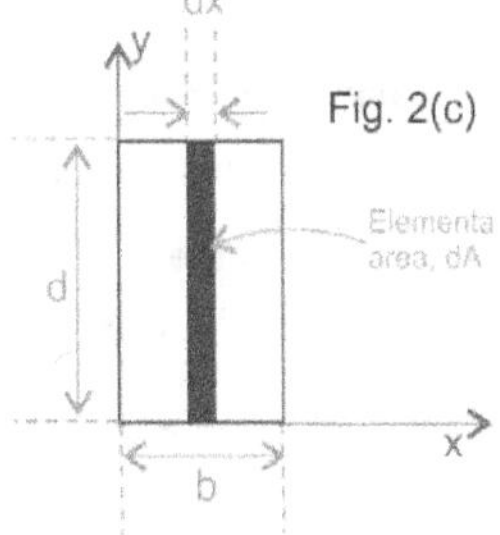

To find I_y :- Consider elemental area dA as shown in Fig 2(c). we see that $dA = d(dx)$. $\therefore I_y = \int_0^b x^2 . b(dy) = \frac{db^3}{3}$

Example Moment of Inertia of Rectangular Lamina about Centroid Axes

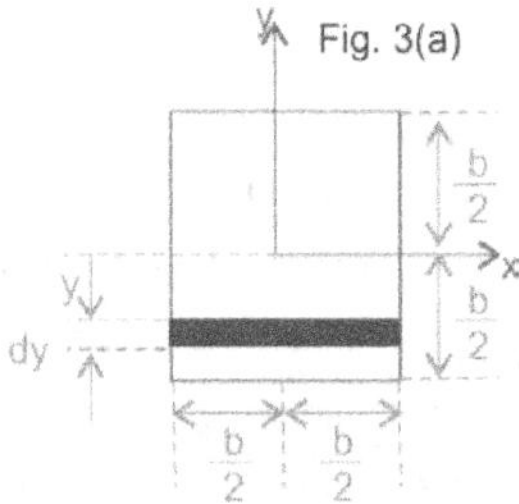

To find I_x: Consider elemental area as shown in the Figure 3(a). $dA = b(dy)$

$$\therefore \bar{I_x} = \int_{-\frac{d}{2}}^{+\frac{d}{2}} y^2 . b . (dy) = b\left[\frac{y^3}{3}\right]_{-\frac{d}{2}}^{\frac{d}{2}} = b\left[\frac{d^3}{24} + \frac{d^3}{24}\right] = \frac{bd^3}{12}$$

Similarly prove that $\bar{I_y} = \frac{db^3}{12}$

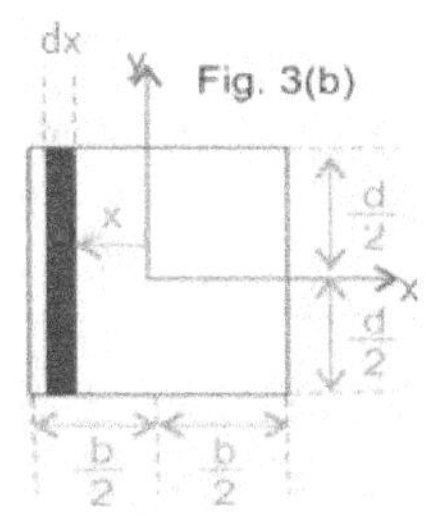

NOTE - I_x and I_y computed in Example 1 are MI about x & y. I_x and I_y computed in Example 2 are MI about x and y axis passing through centroid of rectangle.

Note about Moment of Inertia

- I_x and I_y in first example are MI about x and y axes.
- I_x and I_y in second example are MI about x and y axes passing through centroid of rectangle.
- Did you notice any relationship between these two?

1.14. Transfer Axis Formula

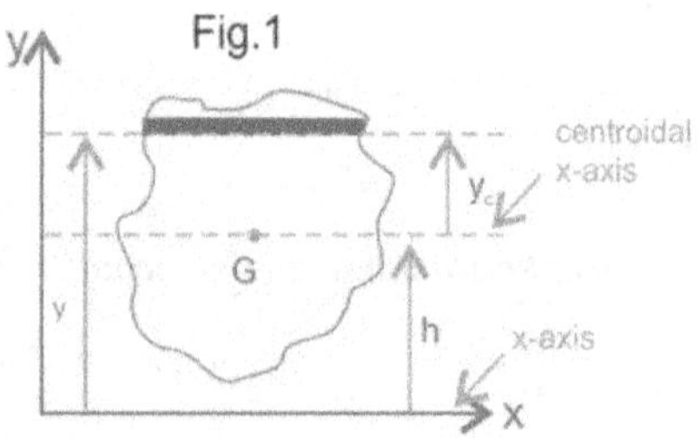

Very often we come across a situation in which we know moment of inertia about centroidal axis of an object and we need to compute moment of inertia about some other axis which is parallel to the centroidal axis. In such situations, we use transfer axis formula also called Parallel axis theorem some times. Before we present the Transfer Axis Formula and prove it, we will explore one important and obvious fact about centroid.

Lemma: Centroidal Axis Integral

Lemma: Consider a plane geometric figure shown in Figure 1 at a distance of yc from centroidal x-axis. Then the integral $\int_A y_c(dA) = 0$ where the integral is evaluated over the entire area.

Proof

To compute the distance of centroid from x-axis (shown by the distance h, the y-coordinate of G), we use the formula,

$$h = \frac{\int y(dA)}{\int(dA)} = \frac{\int y(dA)}{A} \Rightarrow hA = \int y(dA) \quad (1)$$

Again, from Fig 1, we see that, $y = y_c + h$ where y_c is the distance of the same element from centroidal x-axis which is parallel to x-axis. Substitute this value of y into equation (1) to get,

$$hA = \int_A y(dA)$$
$$\Rightarrow hA = \int_A (y_c + h)dA$$
$$\Rightarrow hA = \int_A y_c(dA) + hA \qquad (2)$$
$$\Rightarrow \int_A y_c(dA) = 0$$

Carefully note that (dA) is elemental area and y_c is distance from centroidal axis. We use this result in proving Transfer axis theorem.

Transfer Axis Formula

If X_c is centroidal axis parallel to x-axis,

$I_x = I_{xc} + h^2 A$ where $\qquad I_x = MI$ w.r.t. x-axis

$\qquad\qquad\qquad\qquad\qquad I_{xc} = MI$ w.r.t. centroidal x-axis,

$\qquad\qquad\qquad\qquad\qquad h =$ distance between X and X_c axes,

$\qquad\qquad$ and $\qquad A =$ Area of the geometric figure

Fig.2: $y = y_c + h$

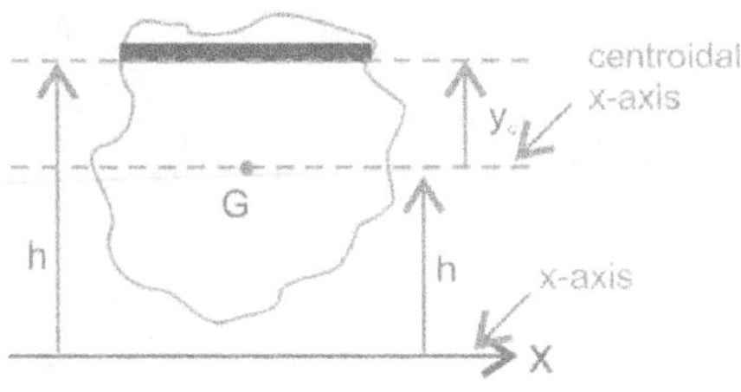

Proof:- Consider Fig 2 where a geometric shape of area "A" is shown. Also shown in Fig 2 are two axes: (a) x-axis and (b) centroidal axis X_c which is parallel to X. We denote the centroidal axis as X_c. By definition of moment of inertia, we have,

$$I_x = \int_A y^2(dA) \ \text{------(3)}$$
$$\text{and } I_{xc} = \int_A y_c^2(dA) \ \text{-----(4)}$$

now, Eqn(3) $\Rightarrow \int_A y^2(dA) = \int_A (y_c + h)^2(dA) = \int_A (y_c^2 + h^2 + 2hy_c)(dA)$

$$\underbrace{\int_A y_c^2(dA)}_{\text{see Eqn(4)}} + h^2 \int (dA) + \underbrace{2h \int_A y_c(dA)}_{\text{see Eqn(2)}}$$

Or

$$\boxed{I_x = I_{xc} + h^2 A}$$

Transfer Formula finds its use in computing the moment of inertia of a composite geometric figure with respect to a given axis. We will illustrate this with a couple of examples. Before that, it is very apt to list a few MI values here.

Note carefully that both M.I. w.r.t. x and centroidal x-axis are given below:-

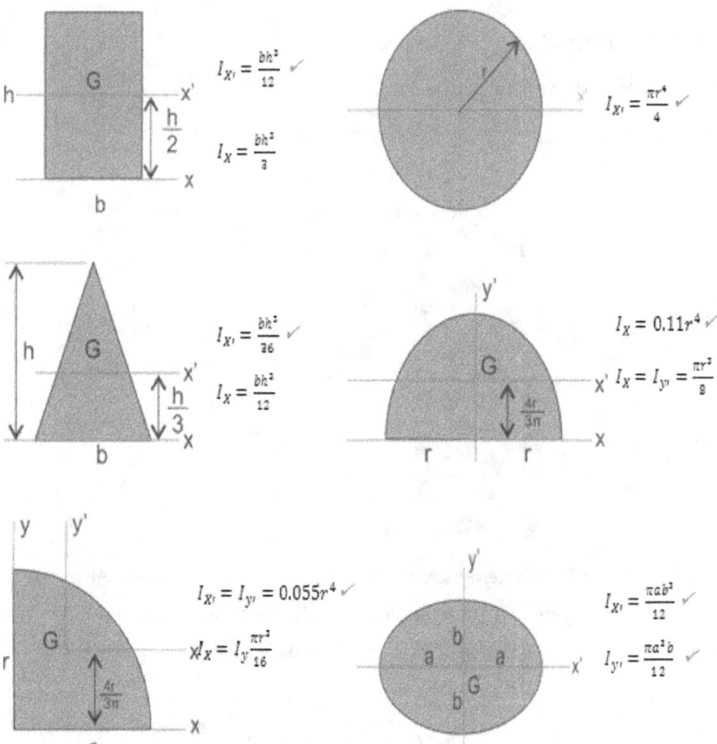

Students are encouraged to derive all the above formulae and convince themselves of the correctness. In specific students need to memorize the formula indicated by the pink tick-mark (✓) above. These values represent moments of inertia with respect to centroidal axis. We will be using these values a lot and I am sure you will be happy that you memorized.

A few examples are worked out here and the students are encouraged to work through these problems independently.

Example 1: Find the Mi of shape shown in Figure 1 with respect to the xx axis shown in Figure 3.

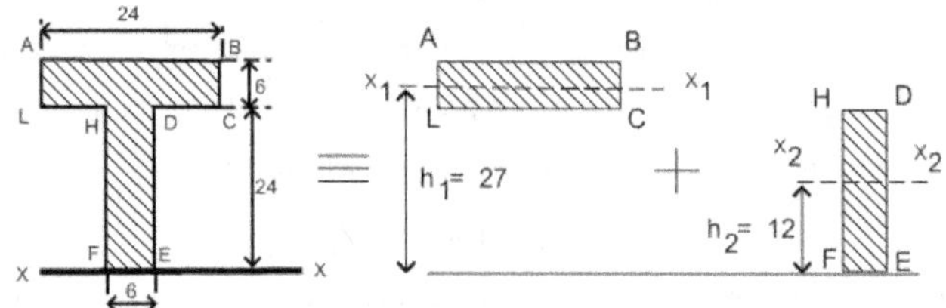

MI of ABCL about its own axis = $I_{x1} = \frac{1}{12}(24)(6)^3 = 432\ mm^2$ and

MI of DEFH about its own axis = $I_{x2} = \frac{1}{12}(6)(24)^3 = 6912\ mm^2$

$$I_x = \left\{\begin{matrix} MI\ of\ T - section \\ w.r.t.xx - axis \end{matrix}\right\} = \overbrace{\left\{\begin{matrix} MI\ of\ ABCL \\ w.r.t.xx - axis \end{matrix}\right\}}^{I_{x1}+h_1^2 A_1} + \overbrace{\left\{\begin{matrix} MI\ of\ DEFH \\ w.r.t.xx - axis \end{matrix}\right\}}^{I_{x2}+h_2^2 A_2}$$

$= \quad 432 + (27)^2[6\text{x}24] + 6912 + (12)^2[6\text{x}24]$

$= \quad 105408 + 27648$

$= \quad 133056\ mm^4$

It is often useful to Tabulate the calculations as shown for compactness and systematic presentation of information

Sl.No.	Object	Area A_i	I_{xi}	h_i	$I_x = I_{xi} + h_i^2 A_i$
1	ABCL	144 mm²	$\frac{24 * 6^3}{12}$	27	$432 + 27^2(6 * 24) = 105408\ mm^4$
2	DEFH	144 mm²	$\frac{6 * 24^3}{12}$	12	$6912 + 12^2(6 * 24) = 27648\ mm^4$
					$133056\ mm^4$

1.15. Mass Moment of Inertia

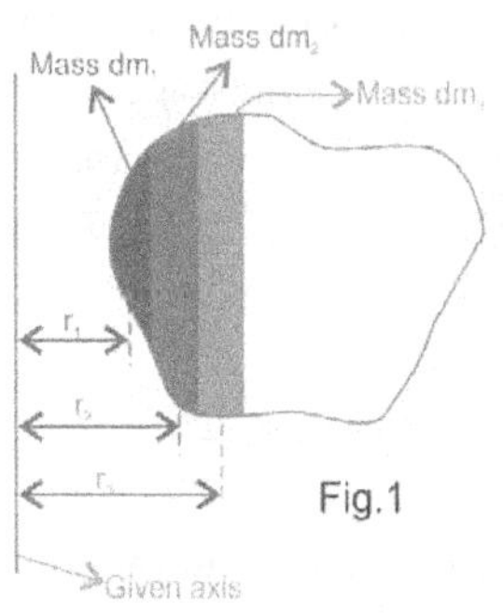

Product of mass and the square of the distance of the center of gravity of the mass from an axis is known as mass moment of inertia about that axis. Mass moment of inertia about the x-axis is represented by I_{mx} and about y-ax is represented by I_{my}. Consider the body shown in Fig. 1 the mass moment of inertia about the given axis is defined as:

$$I_m = r_1^2(dm_1) + r_2^2(dm_2) + r_3^2(dm_3) + \cdots \ldots \ldots \ldots \ldots$$

Where dm_1, dm_2, dm_3 are elemental masses and r_1, r_2, r_3 etc. are distances of elemental masses from given axis. If the total volume of the body is "spanned" by infinitely many, infinitely small elements, the discrete infinite summation may be written as:

$$I_m = \int_v r^2 \, (dm)$$

The term $r^2 dm$ was first used by Christiaan Huygens in 1673 in his study on Compound Pendulum and the term moment of inertia was introduced by Leonhard Euler in 1765. This term is useful in developing equations of motion of a body undergoing rotary motion.

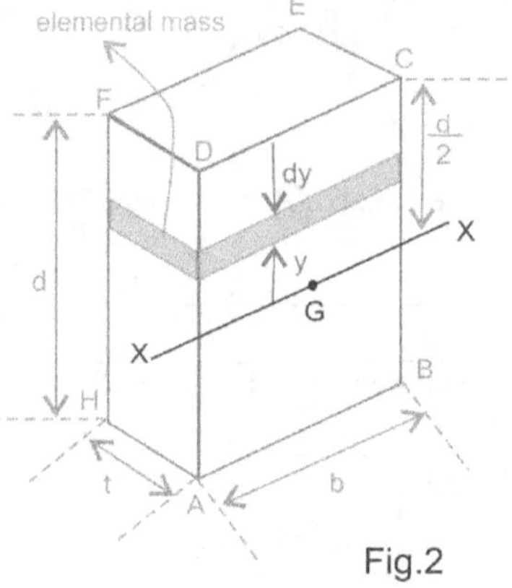

Fig.2

Example: Mass MI of Parallelopiped

Let M be mass, ρ be density of material and with dimensions shown in Fig.2. Consider centroidal x-axis passing through CG of the body G as shown in Figure 2. Consider elemental mass dm as shown by a small slice of the plate parallel to the plane CEFD and at a distance of y from xx axis and having thickness "dy". Mass dm may be computed as:

$dm = b(dy)t \times \rho$

Therefore, the elemental inertia may be written as:

$$dI_{mx} = y^2(dm) = y^2 bt \, \rho(dy)$$

and the total inertia may be obtained by integrating above equation from $\frac{-d}{2}$ to $\frac{d}{2}$:

$$I_{mx} = \int_{-\frac{d}{2}}^{\frac{d}{2}} dI_{m_x} = \rho bt \int_{-d}^{\frac{d}{2}} y^2 (dy) = \frac{\rho bt}{3}\left[\frac{d^3}{8} + \frac{d^3}{8}\right] = \frac{\rho bt}{3} * \frac{2d^3}{8} = \frac{bd^3}{12} * \rho t = I_{xx}\rho t$$

where I_{xx} is the area moment of inertia derived earlier. It may be noted that

$$I_{mx} = \frac{\rho bt\, b^3}{12} = \frac{1}{12}(\rho btd)d^2 = \frac{1}{12}Md^2$$

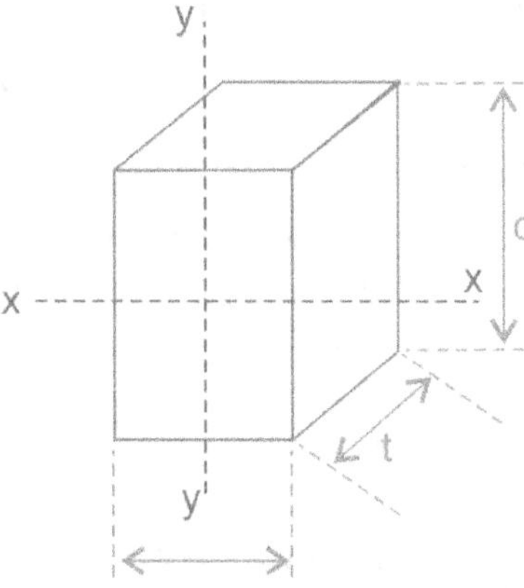

Similarly, mass moment of inertia of the rectangular plate about yy axis passing through the CG of the plate is obtained as: $I_{my} = \frac{1}{12}Mb^2$ students are encouraged to derive this.

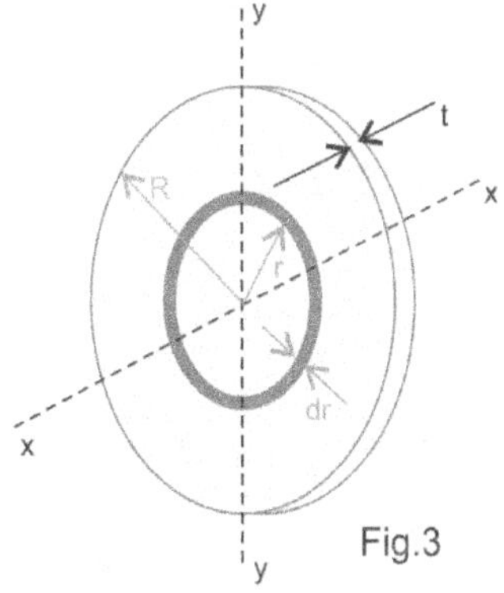

Fig.3

Example: Mass MI of Circular Disc

In figure note that "t" is the thickness of the disc perpendicular to the $x - y$ plane and R is the radius of the disc. Consider an elementary circular ring of radius r and thickness dr as shown in Fig.3. The area of the ring, $dA = 2\pi r(dr) \Rightarrow volume\ of\ ring = 2\pi r(dr)t \Rightarrow dm = mass\ of\ ring = \rho 2\pi r(dr)t$. Carefully note that r is the distance of the elemental ring from

$z-$ axis which is perpendicular to x and y axes. Hence the moment of inertia we are attempting to compute is about $z-$ axis. The elemental moment of inertia is:

$$dI_{mz} = r^2(dm) = 2\pi\rho t r^3 (dm)$$

$$\Rightarrow I_{mz} = 2\pi\rho t \int_0^R r^3 (dr) = 2\pi\rho t \frac{R^4}{4} = \pi R^2 \rho t. \frac{R^2}{2} = \frac{1}{2} MR^2$$

We know that $I_{mx} + I_{my} = I_{mz}$ and $I_{mx} = I_{my}$ by symmetry. $\therefore I_{mx} = I_{my} = \frac{1}{4} MR^2$

Parallel Axis Theorem for Mass MI

It may be noted that parallel axis theorem applies to Mass Moment of inertia also. If I_{mx} is mass moment of inertia of a body about $x-$axis, I_{mxG} is mass moment of inertia about an axis parallel to $x-$axis and passing through CG, and h is the distance between the axes then:

$I_{mx} = I_{mxG} + Mh^2.$

1. *Kinematics*

We recall that the study of mechanics is divided into two broad branches viz., (i) Statics, and (ii) Dynamics. We have thus far seen aspects of statics. The branch of dynamics as shown in the schematic below is developed by Physicists and Mathematicians such as Galioleo, Newton, Euler, and Lagrange.

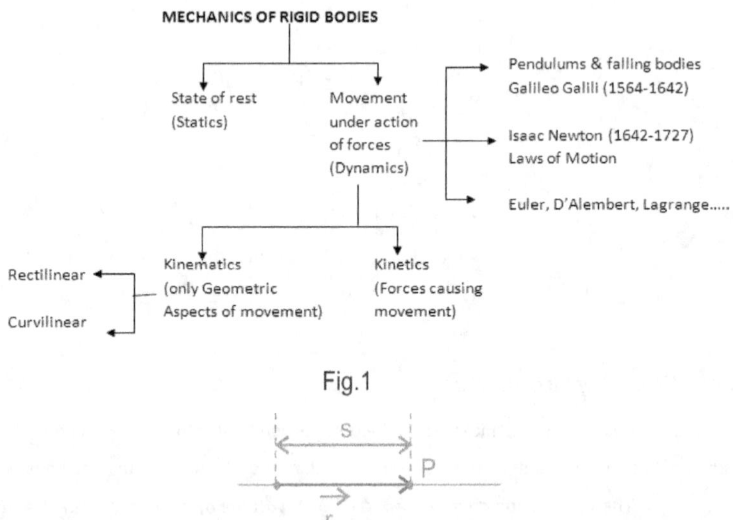

Fig.1

We study the branches of Dynamics, viz., Kinematics and Kinetics in the following.

- ### *Rectilinear Kinematics*

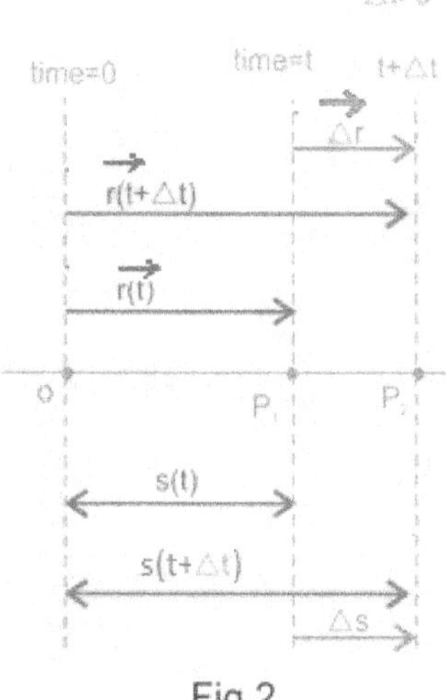

Fig.2

When the motion of the body can be characterized by motion of its mass center and any rotation of the body can be neglected, we study kinematics of "particle". We begin study of kinematics with motion along a straight line by studying about position, velocity and acceleration.

Position

Position of Particle P <u>at a given instant</u> may be specified by a single coordinates "s" or by position vector, $\vec{r}$, along s-axis with respect to origin O., "s" is considered positive if the particle is to the right of O and negative if it is to the left.

Displacement

Displacement is change in position. In Fig.2, Particle P, initially at origin O at time = 0, is displaced to position P_1 at time = t and further to P_2 at time = $t + \Delta t$ $(\Delta t > 0)$

In Figure 2, both $\vec{r}(t)$ and $\vec{r}(t + \Delta t)$ are along s-axis. The displacement during time internal $(t, t + \Delta t)$ is $\Delta \vec{r}$ which is again along s-axis. We write $\overrightarrow{\Delta r} = \vec{r}(t + \Delta t) - \vec{r}(t)$. Similarly, we can write $\Delta s = s(t + \Delta t) - s(t)$. If P_2 is to the right of P_1, $\Delta s > 0$ else $\Delta s \leq 0$. Note that $\overrightarrow{\Delta r}$, displacement is a vector quantity and is different from the positive scalar which is commonly termed "distance traveled" which represents total length of path over which the particle travels.

Velocity

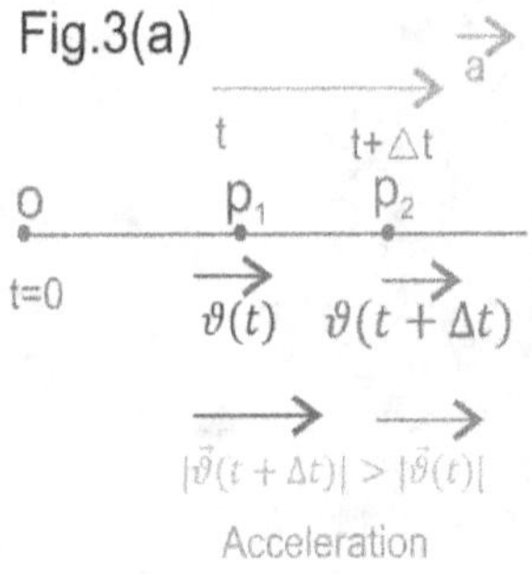

Fig.3(a)

$\vec{a}$

$t \qquad t+\triangle t$

O $\qquad$ P$_1$ $\qquad$ P$_2$

$t=0$

$\overrightarrow{\vartheta(t)} \quad \overrightarrow{\vartheta(t+\Delta t)}$

$|\vec{\vartheta}(t+\Delta t)| > |\vec{\vartheta}(t)|$

Acceleration

We define the quantities average velocity, instantaneous velocity, speed and average speed. Referring to Fig 2, average velocity of the particle as it travels from P$_1$ to P$_2$ <u>during the time interval from t to $(t+\Delta t)$</u> is defined as $\overrightarrow{v_{avg}} = \frac{\Delta \vec{r}}{\Delta t}$. If we consider the limiting case of $\Delta t \to 0$, the instantaneous velocity is defined as $\vec{\vartheta} = \lim_{\Delta t \to 0} \frac{\Delta \vec{r}}{\Delta t} = \frac{d\vec{r}}{dt}$. Note that $\overrightarrow{\vartheta_{avg}}$ and $\vec{\vartheta}$ both are vector quantities. The magnitude of velocity is called speed: **$speed = |\vec{\vartheta}|$. Average speed** is always a positive scalar and is defined as the <u>total distance</u> travelled divided by the elapsed time:

$$\vartheta_{avg \,.speed} = \frac{S_T}{\Delta t}$$

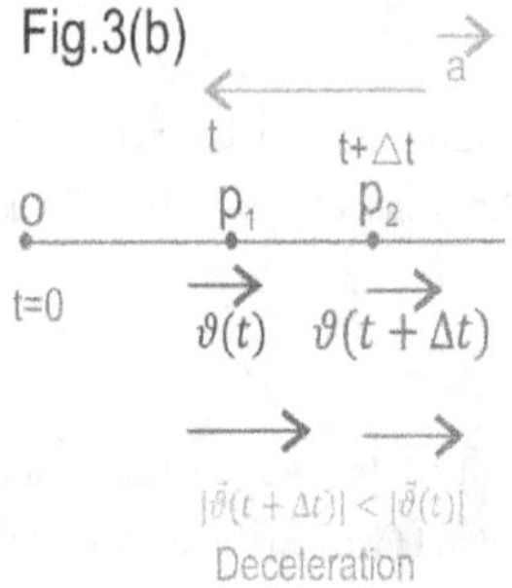

Fig.3(b)

$\vec{a}$

$t \qquad t+\triangle t$

O $\qquad$ P$_1$ $\qquad$ P$_2$

$t=0$

$\overrightarrow{\vartheta(t)} \quad \overrightarrow{\vartheta(t+\Delta t)}$

$|\vec{\vartheta}(t+\Delta t)| < |\vec{\vartheta}(t)|$

Deceleration

Acceleration

Provided that velocities of particle at time instants t & $(t+\Delta t)$ are known, average acceleration as it moves from P$_1$ to P$_2$ is defined as

$$\overrightarrow{a_{avg}} = \frac{\Delta \vec{\vartheta}}{\Delta t} = \frac{\vec{\vartheta}(t+\Delta t) - \vec{\vartheta}(t)}{\Delta t} = \frac{\Delta \vec{\vartheta}}{\Delta t}$$

And the **instantaneous acceleration** at time t is defined as:

$$\vec{a} = \lim_{\Delta t \to 0} \frac{\Delta \vec{\vartheta}}{\Delta t} = \frac{\Delta \vec{\vartheta}}{dt}$$

If the velocity is increasing, i.e. $\left|\vec{\vartheta}(t + \Delta t)\right| > \left|\vec{\vartheta}(t)\right|$, then we say that the particle is +accelerating {see Fig.3(a)}. If $\left|\vec{\vartheta}(t + \Delta t)\right| < \left|\vec{\vartheta}(t)\right|$ then we say that the particle is decelerating {see Fig.3(b)}. Often the vector equations we considered so fat in defining velocity may also be written as:

$$Eqn.(a) \qquad\qquad\qquad\qquad Eqn.(b)$$
$$\vartheta = \frac{ds}{dt} \qquad \text{and} \qquad a = \frac{d\vartheta}{dt} = \frac{d^2 s}{dt^2}$$

From (a) and (b), we see that $a = \dfrac{d\vartheta}{dt} = \dfrac{d\vartheta}{ds} \cdot \dfrac{ds}{dt} = \vartheta \dfrac{d\vartheta}{ds}$ and $a = \vartheta \dfrac{d\vartheta}{ds} \Rightarrow \overline{a(ds) = \vartheta(d\vartheta)}$ $Eqn.(c)$

If acceleration is constant, using (a), (b) and (c) we can write formulae that relate a, ϑ, s and t.

Constant Acceleration

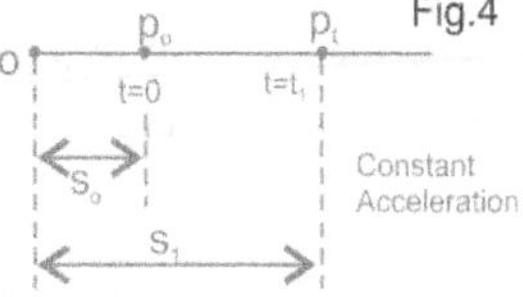

Conditions in Fig.4 may be written as:

At time $t = 0, s = s_0, \vartheta = \vartheta_0, a = a_c$

At time $t = t_1, s = s_1, \vartheta = \vartheta_1, a = a_c$

(a)$\Rightarrow \vartheta_1 = \vartheta_0 + a_c t_1$ \qquad (b)$\Rightarrow s_1 = s_0 + \vartheta_0 t_1 + \frac{1}{2} a_c t_1^2$ \qquad (c)$\Rightarrow \vartheta_1^2 - \vartheta_0^2 = 2a_c(s_1 - s_0)$

Remember that the last three equations are valid only if acceleration is constant.

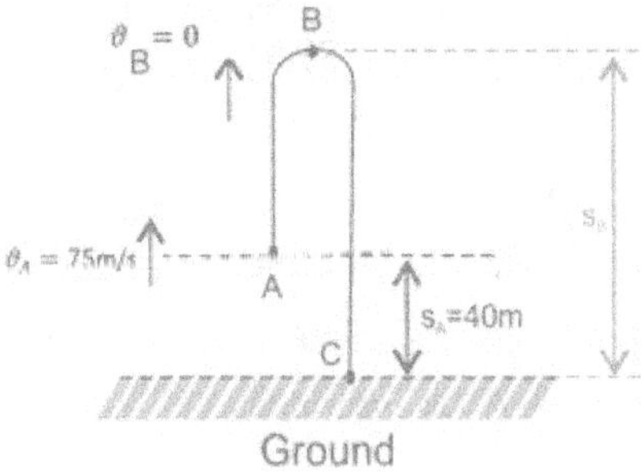

Example: Rocket Problem

During a test, a rocket is traveling upward at 75 m/s, and when it is 40 m from ground, its engine fails. Determine the maximum height reached by the rocket and its speed just before it hits the ground. While in motion, the rocket is subjected to a constant downward acceleration of 9.81 m/s² due to gravity. Neglect the effort of air resistance.

This problem is constant acceleration problem since acceleration due to gravity

$$g = a_c = -9.81 \text{ m/s}^2 \text{ always.}$$

Between points $A\&B$: $- \vartheta_B^2 = \vartheta_A^2 + 2a_c(s_B - s_A)$

$$\Rightarrow 0^2 = \left(\tfrac{75m}{s}\right)^2 + 2\left(-9.81\tfrac{m}{s^2}\right)(s_B - 40m)$$

$$\Rightarrow s_B = 327m$$

Between points $A\&C$: $- \vartheta_B^2 = \vartheta_A^2 + 2a_c(s_c - s_A) = 75^2 + 2(-9 - 81)(0 - 40)$

$$\Rightarrow \vartheta_0 = -80.1m/s \quad [\text{downward}]$$

Example: Variable Acceleration: Movement in Viscous Medium

A small projectile is fired downward into a fluid medium with an initial velocity of 60 m/s. due to the resistance of the fluid, the projectile experiences a deceleration equal to $a = (-0.4\vartheta^3)$m/s², where ϑ is projectiles velocity in m/s. determine the projectiles velocity and position 4 second after it is fired.

Solution: Since the motion of projectile is downward, consider downward velocities and acceleration as positive – Note carefully that acceleration is NOT constant. We know that at $t = 0$, $\vartheta = 60$m/s. Let $\vartheta = \vartheta_1$ at $t = t$sec with this information, we not that

$$a = \frac{d\vartheta}{dt} = -0.4\vartheta^3 \Rightarrow \int_{60}^{\vartheta_1} \frac{d\vartheta}{-0.4\vartheta^3} = \int_0^t dt \Rightarrow \vartheta_1 = \sqrt{\frac{1}{\frac{1}{60^2}+0.8t}} \quad \therefore when\ t = 4sec, \vartheta = 0.559\text{m/s}$$

[downward]

$$\vartheta = \frac{ds}{dt}\sqrt{\frac{1}{\frac{1}{60^2}+0.8t}} \Rightarrow \int_0^t \sqrt{\frac{1}{\frac{1}{60^2}+0.8t}} \Rightarrow s = \frac{1}{0.4}\left\{\sqrt{\frac{1}{60^2}+0.8t} - \frac{1}{60}\right\} \quad \therefore s = 4.43m\ at\ t = 4sec$$

1.16. Curvilinear Motion

In this session, the following topics are presented:

1) General Curvilinear motion (position, velocity, acceleration).
2) Curvilinear motion in Rectangular components.
3) Motion of a projectile.

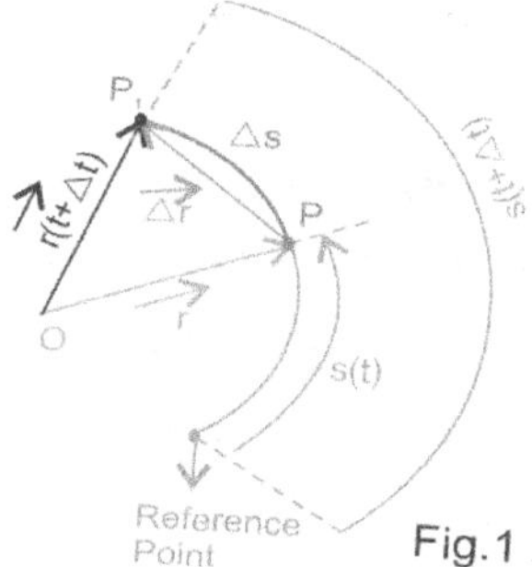

Fig.1

General Curvilinear Motion

Consider movement of a particle P along path as shown in Figure 1. The position of the point at time t, indicated by P, can be described by distance travelled by P along the curve from a reference point on the curve shown as "s" in Fig. 1.

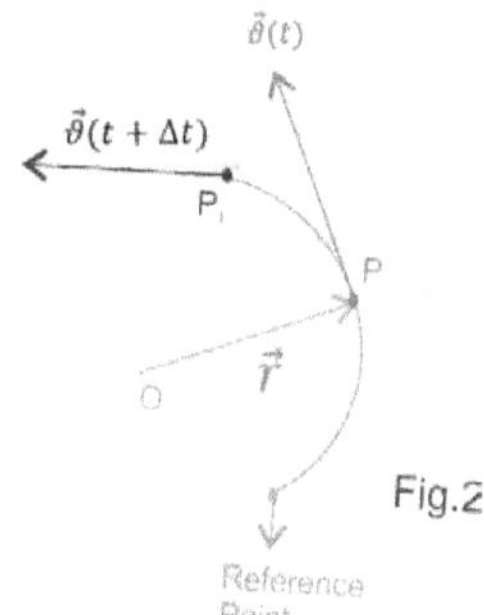

Fig.2

Alternatively, an origin O may be chosen suitably and a <u>position vector</u>, $\vec{r}(t)$, may be used to describe the <u>position</u> of P at time t. When point P moves to a new location P_1, the <u>change in position</u> may be described using movement along the curve, shown in **pink color** or by chord shown in **green color** in Figure 1.

The chord shown in **green** colour, $\overrightarrow{\Delta r}$, is termed displacement to represent change in particle's position and we write, $\Delta\vec{r} = \vec{r}(t + \Delta t) - \vec{r}(t)$. Average velocity of the particle, $\vec{\vartheta}_{avg}$, is defined as $\vec{\vartheta}_{avg} = \frac{\Delta\vec{r}}{\Delta t}$ <u>during $(t, t + \Delta t)$</u>. Instantaneous velocity is obtained when $\Delta t \to 0$. $\vec{\vartheta} = \lim_{\Delta t \to 0} \frac{\overrightarrow{\Delta r}}{\Delta t}$. we see that length of $\overrightarrow{\Delta r}$ and length of Δs become closer and closer as $\Delta t \to 0$. So.

The magnitude of $\vec{\vartheta}$, denoted by v, without arrow on head may be obtained as:

$v = |\vec{\vartheta}| = \lim_{\Delta t \to 0} \frac{\Delta s}{\Delta t} = \frac{ds}{dt}$. Velocity vector $\vec{\vartheta}$, at time instant t is shown in Fig 2. Notice that $\vec{\vartheta}$ is tangent to path at <u>time t</u>.

Similarly we can construct velocity vector at time instant $(t + \Delta t)$ by drawing tangent to the path at particles new position P_1. Average acceleration during the time interval from t to $t + \Delta t$ may be noted as $\vec{a}_{avg} = \frac{\Delta \vec{\vartheta}}{\Delta t} = \frac{\vec{\vartheta}(t+\Delta t) - \vec{\vartheta}(t)}{\Delta t}$. To describe acceleration in a graphical representation, consider Figure 3 where both $\vec{\vartheta}(t + \Delta t)$ and $\vec{\vartheta}$ are drawn with their tails coinciding at a point O_1.

The arrow-heads, when joined by a smooth curve, are said to trace a <u>hodograph</u>. Further we see that the chord of the hodograph, $\Delta \vec{\vartheta} = \vec{\vartheta}(t + \Delta t) - \vartheta(t)$.

Thus, <u>average acceleration</u> may be defined as $\vec{a}_{avg} = \frac{\Delta \vec{\vartheta}}{\Delta t}$ <u>during time interval $(t, t + \Delta t)$</u> and instantaneous acceleration may be defined as $\vec{a} = \lim_{\Delta t \to 0} \frac{\Delta \vec{\vartheta}}{\Delta t} = \frac{d\vec{\vartheta}}{dt}$. Note that Acceleration is tangential to hodograph.

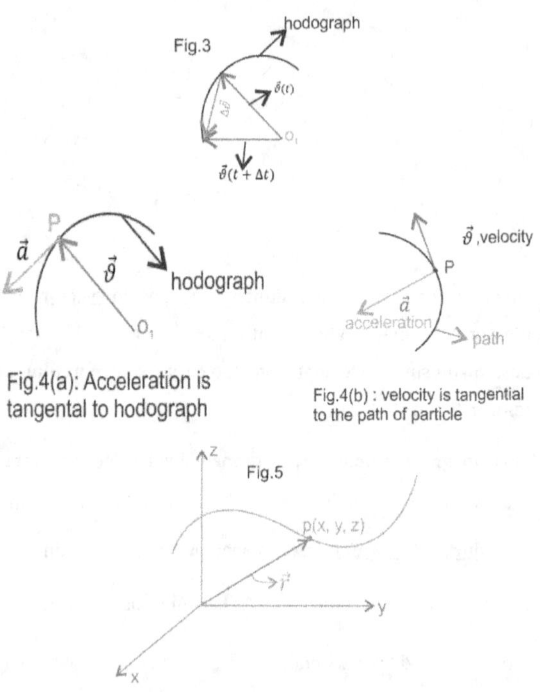

Fig.4(a): Acceleration is tangental to hodograph

Fig.4(b) : velocity is tangential to the path of particle

Rectangular Components

Consider Point P on space curve as shown in Fig 5. Let (x, y, z) be coordinates of point P. As P moves along the curve, x, y, z change. The position vector of P is given by

$\vec{r}(t) = x(t)\vec{\imath} + y(t)\vec{\jmath} + z(t)\vec{k}$ and its magnitude is obtained by

$$r = \sqrt{x^2 + y^2 + z^2}.$$

Similarly, its velocity is obtained as $\vec{\vartheta} = \dfrac{d\vec{r}}{dt} = \dfrac{dx}{dt}\vec{\imath} + \dfrac{dy}{dt}\vec{\jmath} + \dfrac{dz}{dt}\vec{k}$ or $\vec{\vartheta} = \vartheta_x\vec{\imath} + \vartheta_y\vec{\jmath} + \vartheta_z\vec{k}$ where ϑ_x, ϑ_y and ϑ_z are scalar components of $\vec{\vartheta}$ along x, y, z – axes, where

$$\vartheta_x = \frac{dx}{dt}, \vartheta_y = \frac{dy}{dt}, \vartheta_z = \frac{dz}{dt}.$$

Magnitude of velocity $\vartheta = |\vec{\vartheta}| = \sqrt{\vartheta_x^2 + \vartheta_y^2 + \vartheta_z^2}.$

Similarly, acceleration and magnitude of acceleration may be given by

$$\vec{a} = a_x\vec{\imath} + a_y\vec{\jmath} + a_z\vec{k}, a_x = \frac{d\vartheta_x}{dt} = x, a_y = \frac{d\vartheta_y}{dt} = y, a_z = \frac{dz}{dt} = z, \text{ and } a = |\vec{a}| = \sqrt{a_x^2 + a_y^2 + a_z^2}.$$

This idea can be applied effectively to the analysis of movement of an object which is thrown with an initial velocity. Such analysis is often performed under the assumption that air resistance is negligible. With such an assumption, the projectile can be considered to move in a single plane, xy – plane

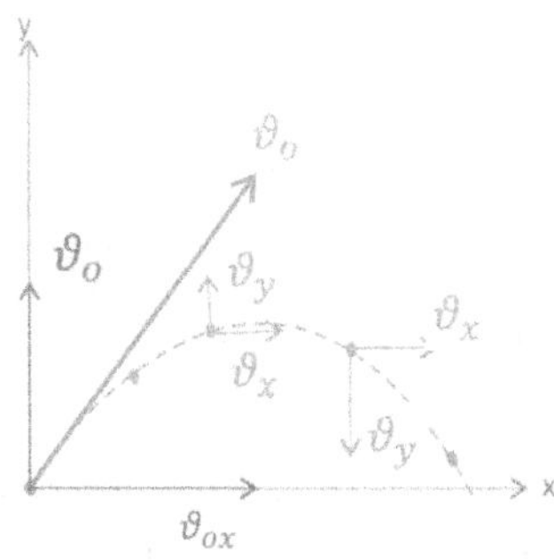

Motion of a Projectile

Consider a particle location at origin. At time t = 0, the particle has a displacement of zero along x and y – axes. Further, at this very instant, the particle is given an initial velocity of ϑ_0. This initial velocity has an x –component of ϑ_{0x} and a y –component of ϑ_{0y}. The kinematic analysis of the particle may be carried along x –axis and y –axis separately. The kinematic details of the particle may now be written as:

<table>
<tr><td>

Along x −axis

At $t = 0, u_o = \vartheta_{ox}, S_o = 0$

Since there is no force along x −axis, $a_x = 0$ for all $t > 0$. So the kinematic equations for constant acceleration apply.

$S_x = \vartheta_{ox} t$

$\vartheta_x = \vartheta_{ox}$

</td><td>

Along y −axis

At $t = 0, u_o = \vartheta_{oy}, S_o = 0$

Since there is constant gravitation force, there is constant downward acceleration. Hence, $a_y = -g$ for all $t > 0$. So, constant acceleration equation apply here also

$S_y = \vartheta_{oy} t - \frac{1}{2} g t^2$

$\vartheta_y = \vartheta_{oy} t - g t$

$\vartheta_y^2 - \vartheta_{oy}^2 = 2 a_y S_y$

</td></tr>
</table>

We will see few problems on projectile motion.

- ***Example: Range and Maximum Height of Projectile***

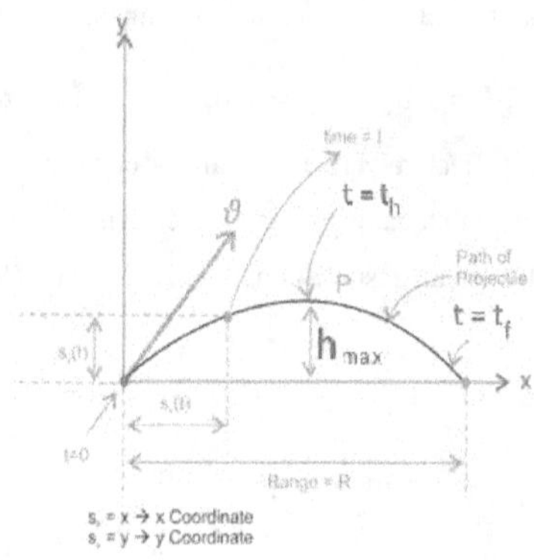

A projectile is fired at an angle of $\propto$ from level ground with a velocity if ϑ. Find its range and max. Height reached above ground.

Solution: First step is to establish x, y −axes. Choose the location of origin as shown in Figure below. The x −component of ϑ is $\vartheta \cos \alpha$ and its y −component is $\vartheta \sin \alpha$. The projectile starts its flight at $t = 0$, attains max height at $t = t_h$ and ends its flight at time $t = t_f$.

<u>Along $x - axis$:</u>

$a_x(t) = o \ \forall t > 0$

$\vartheta_x(t) = \vartheta \cos \alpha \ \ \forall t > 0$

$S_x(t) = (\vartheta \cos \alpha)t = x \Rightarrow t = \dfrac{x}{\vartheta \cos \alpha} \cdots\cdots (A)$

<u>Along $y - axis$</u>

$a_y(t) = -g \ \forall t > 0$

$$\vartheta_y(t) = \vartheta \sin \alpha - gt$$

$$s_y(t) = (\vartheta \sin \alpha)t - \frac{1}{2}gt^2 = y \quad \cdots \cdots \text{(B)}$$

$$\therefore y = \vartheta \sin \alpha \frac{x}{\vartheta \cos \alpha} - \frac{1}{2}g\frac{x^2}{\vartheta^2 \cos^2 \alpha} = x \tan \alpha - \frac{1}{2}\frac{gx^2}{\vartheta^2 \cos^2 \alpha}.$$

This is the equation of the projectile's path in xy-plane. At $t=t_f$, the flight of projectile ends and hence $x = R$ and $y=0$. Substitute $x=R$ and $y=0$ in the equation of projectile's path, we get,

$$R = \frac{\vartheta^2 \sin 2\alpha}{g}. \quad \text{Substitute y=0 at t = t}_f \text{ in (B) to get } t_f = \frac{2\vartheta \sin \alpha}{g}.$$

To find the time at which the projectile attains maximum height, h_{max}, notice that the tangent to the projectile's path is horizontal or $\frac{dy}{dx} = 0$ at point P. Thus,

$$\tan \alpha = \frac{1}{2}\frac{2gx}{\vartheta^2 \cos^2 \alpha} \Rightarrow x = \frac{\vartheta^2 \sin(2\alpha)}{2g} \text{ at point P.}$$

We also know that $t=t_h$ at point P. Thus at P, $t=t_h$ and $x = \frac{\vartheta^2 \sin(2\alpha)}{2g}$. Use this in (A) to get

$$t_h = \frac{x}{\vartheta \cos \alpha} = \frac{\vartheta \sin \alpha}{g}.$$

Using this value, h_{max} may be obtained by using $y = h_{max}$ at $t=t_h$ as

$$h_{max} = y(t_h) = (\vartheta \sin \alpha)t_h - \frac{1}{2}gt_h^2 = \frac{\vartheta^2 \sin^2 \alpha}{g}.$$

Work through every single step and convince yourself that all the result are correct. It is very essential for understanding further problems.

- **_Example: Projectile Filed from the Top of a Building_**

A projectile is fired from a building of height h with horizontal velocity ϑ. Find the range on ground.

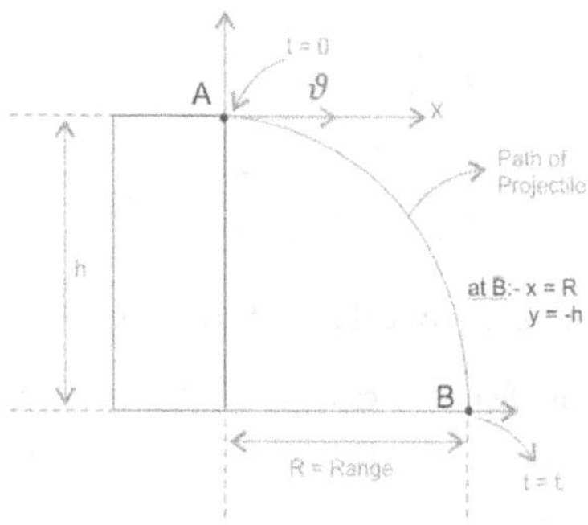

At any point on the path of projectile,

$\vartheta_x = \vartheta,\ a_y = -g.$

Also at $t = 0,\ \vartheta_x(t) = \vartheta$ and $\vartheta_y(t) = 0.$

Along x – axis	Along y –axis
$\vartheta_x(t) = \vartheta\ \forall t > 0$	$\vartheta_y(t) = -gt$
$x = s_x(t) = \vartheta t$	$x = s_y(t) = -\dfrac{1}{2}gt^2$

$$t = \frac{x}{\vartheta} \Rightarrow y - \frac{1}{2}g\frac{x^2}{\vartheta^2}$$

$$\text{At } B: -h = -\frac{1}{2}g\frac{R^2}{\vartheta^2}$$

$$\Rightarrow R = \vartheta\sqrt{\frac{2h}{g}}$$

- ### Example: Computing Time of Flight

A projectile fired from level ground attained a maximum height of 200 m and a range of 1000 m. What is the time of flight of the projectile?

To find the time of flight, we need to compute the angle of firing and the velocity with which the projectile is fired.

$$h_{max} = 200 = \frac{\vartheta^2 \sin^2\alpha}{g} \quad\text{-------- (1)}$$

$$R = 1000 = \frac{\vartheta^2 \sin^2\alpha}{g} \quad\text{----------- (2)}$$

Dividing (1) by (2), we get

$$\frac{200}{1000} = \frac{\sin^2\alpha}{2\sin\alpha\cos\alpha} \Rightarrow \tan\alpha = \frac{4}{10} \Rightarrow \alpha = 21.8^\circ \Rightarrow \vartheta = \sqrt{\frac{200\,g}{\sin^2\alpha}} = 119.3\frac{m}{s}$$

$$\therefore \text{time of flight} = t_f = \frac{2\vartheta\sin\alpha}{g} = 9\ sec$$

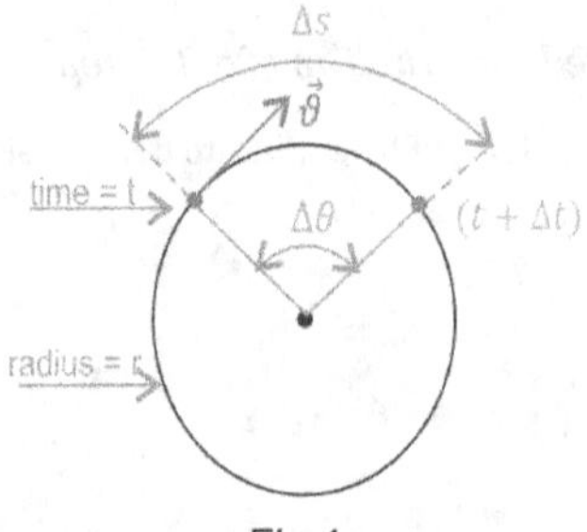

Fig.1

1.17. Tangential and Centripetal Accelerations

Consider a particle moving along a circular path as shown in Figure beside. It is of interest to us to find the magnitude and direction of velocity and acceleration when the particle is at any given point on the circle. In the time interval t to $(t + \Delta t)$, the particle moved a distance of Δs along the arc. Hence, we write an expression for speed as:

Equation 1

$$\vartheta = \lim_{\Delta t \to 0} \frac{\Delta s}{\Delta t} = \lim_{\Delta t \to 0} \frac{r(\Delta \theta)}{\Delta t} = r \lim_{\Delta t \to 0} \frac{\Delta \theta}{\Delta t} = r \frac{d\theta}{dt} = r\omega$$

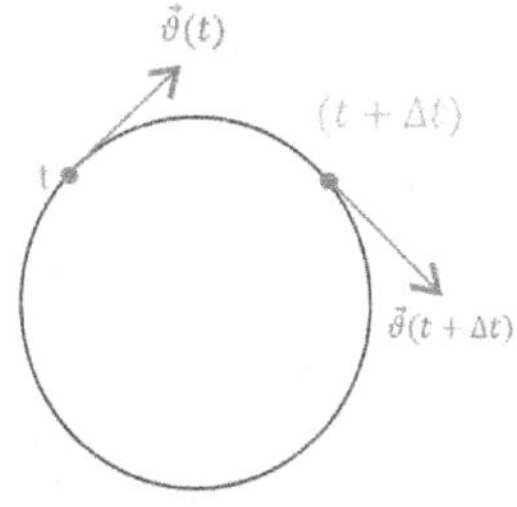

Fig.2

Where ω is angular velocity. It may be noted that $\vec{\vartheta}$, the velocity vector, is tangential to the path at any given instant. Carefully note that $\vec{\vartheta}$ is velocity vector and ϑ is its magnitude so we write: $\vartheta = \frac{ds}{dt} = r\frac{d\theta}{dt} = |\vec{\vartheta}|$.

Again, consider the particle and look at velocities at time instants t and $(t + \Delta t)$. We notice that the velocity vector is changing in two ways: (i) Its magnitude is changing due to tangential acceleration $(\vec{a_t})$ and (ii) direction is changing because of centripetal acceleration $(\vec{a_c})$. The tangential acceleration is tangential to the path and its magnitude is given by

$$a_t = |\vec{a_t}| = \frac{d\vartheta}{dt} = \frac{d}{dt}(r\omega) = r\frac{d\omega}{dt} = r\alpha$$

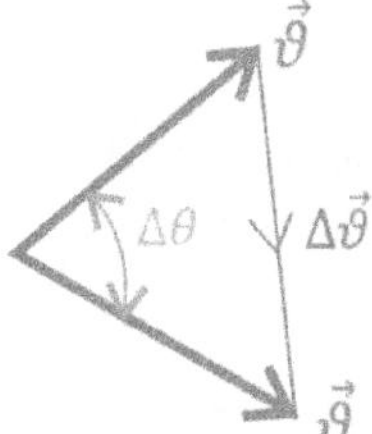

Fig.3: Vector diagram

Where α is angular acceleration. Again, note that ϑ and a_t are magnitudes and $\vec{\vartheta}$ and $\vec{a_t}$ are vectors. Since we have already considered the change in the magnitude through $\vec{a_t}$, we will consider the vector diagram in Fig.3.

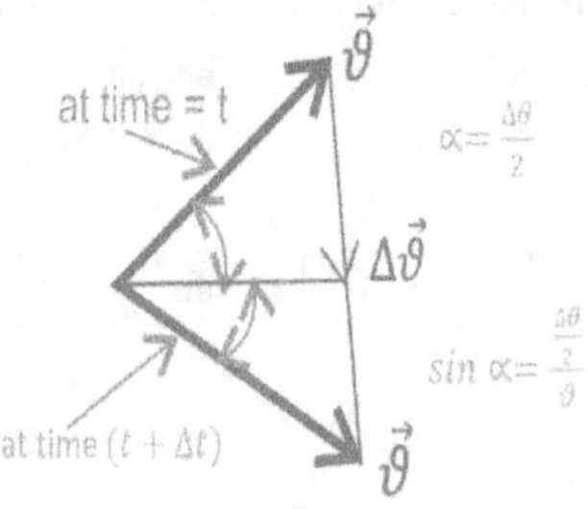

Fig.3: Vector diagram

We notice that the direction of velocity vector from time instant t to that at $(t + \Delta t)$ occurred due to because of $\overrightarrow{\Delta\vartheta}$. So, $(\overrightarrow{a_c})$ may be written as:

$$\overrightarrow{a_c} = \lim_{\Delta t \to 0} \frac{\overrightarrow{\Delta\vartheta}}{\Delta t} = \lim_{\Delta t \to 0} \frac{2\vartheta \sin\left(\frac{\Delta\theta}{2}\right)}{\Delta t}$$

As $\Delta t \to 0, \frac{\Delta\theta}{2} \to 0$ and $\sin\left(\frac{\Delta\theta}{2}\right) \to \frac{\Delta\theta}{2}$. Therefore, we write:

$\longrightarrow$ See Equation 1

$$\overrightarrow{a_c} = \lim_{\Delta t \to 0} \frac{2\vartheta \cdot \frac{\Delta\theta}{2}}{\Delta t} = \vartheta \lim_{\Delta t \to 0} \frac{\Delta\theta}{\Delta t} = \vartheta \frac{d\theta}{dt} = \vartheta \cdot \frac{\vartheta}{r} = \frac{\vartheta^2}{r}$$

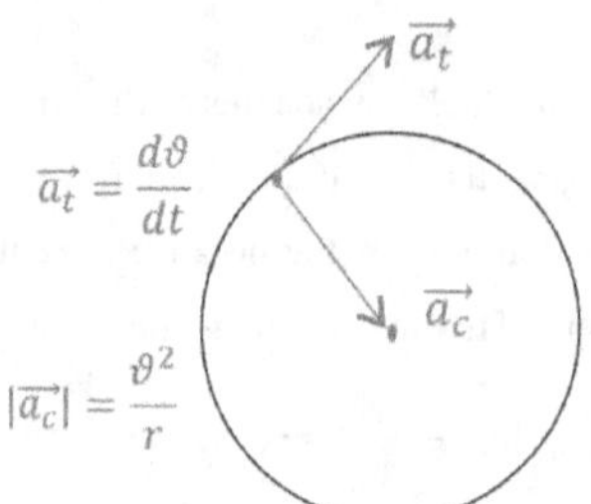

Fig.4: components of acceleration

To sum it up, if a particle is travelling around a circular path and has a velocity of ϑ at time t, its acceleration will have two components:

(i) $\overrightarrow{a_t}$, tangential component with magnitude of $\frac{d\vartheta}{dt}$ and tangential to the path at the instant considered

(ii) $\overrightarrow{a_c}$, centripetal component with magnitude $\frac{\vartheta^2}{r}$ and directed towards the center of the path at the instant considered as shown in Fig.4

Magnitude of total acceleration is given by $|\overrightarrow{a_t} + \overrightarrow{a_n}| = \sqrt{\left(\frac{d\vartheta}{dt}\right)^2 + \left(\frac{\vartheta^2}{r}\right)^2}$

Note

1) If a particle is travelling along a straight line, $r \to \alpha$, hence, it doesn't have centripetal component. It has only $|\overrightarrow{a_t}| = \frac{d\vartheta}{dt}$

2) If the particle travels along circular path with constant "speed", it has only centripetal component $\overrightarrow{a_n}$.

3) If speed of the particle also changes, the total acceleration $= \overrightarrow{a_t} + \overrightarrow{a_c}$

4) The discussion can be applied to any general path $y = f(x)$, by replacing "r" with radius of curvature ρ defined by

$$\rho = \frac{\left[1 + \left(\frac{dy}{dx}\right)^2\right]^{\frac{3}{2}}}{\left|\frac{d^2y}{dx^2}\right|}$$

5) The relation between a_t, t, ϑ and s are the same as the of rectilinear motion:

$a_t = \vartheta, a_t(ds) = \vartheta(d\vartheta)$

If $a_t = $ constant, then, the following equations are applicable:

$s = s_0 + \vartheta_0 t + \frac{1}{2}a_t t^2$

$\vartheta = \vartheta_0 + a_t t$

$\vartheta^2 = \vartheta_0^2 + 2a_t(s - s_0)$

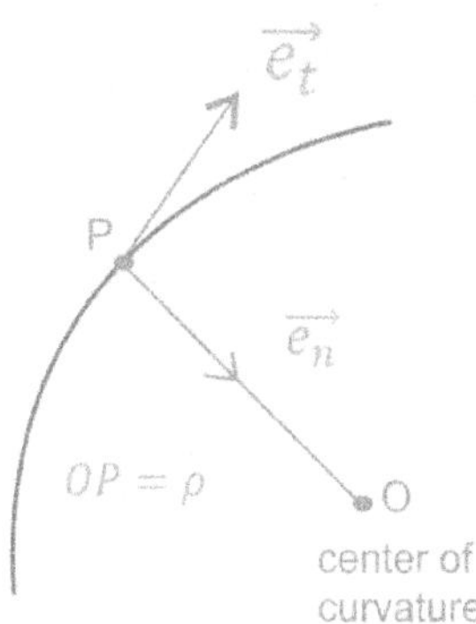

6) If a particle is traveling along a curved path and suppose it is at point P at time instant t. Further suppose $\overrightarrow{e_t}$ is unit vector along tangent at P, $\overrightarrow{e_t}$ is unit vector along normal at P, ϑ is speed ay P, $a_t = \frac{d\vartheta}{dt}$ is the tangential acceleration, and $\rho = $ radius of curvature at P, then, we may write an expression for $\overrightarrow{a_p}$, the total acceleration at P as:

$$\overrightarrow{a_p} = a_t \overrightarrow{e_t} + a_n \overrightarrow{e_n} = \frac{d\vartheta}{dt}\overrightarrow{e_t} + \frac{\vartheta^2}{\rho}\overrightarrow{e_n} \cdots (C)$$

7) This type of analysis is very handy when the path along which the particle moves is known before hand.

Example: Car Travelling through a Dip in the Road

A car passes through a dip in the road at A with constant speed giving it an acceleration of 0.5g. The radius of curvature at A, ρ_A, is 100 m and the distance from the road to the CG of the car, G, is 0.6m. Determine the speed ϑ of the car at A.

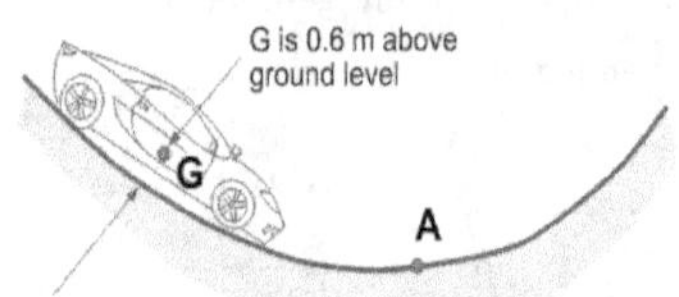

Solution: We treat the car as a particle since the other dimensions are very big as compared to the dimensions of the car. Hence we compute the radius of curvature of the path traced by the centre of gravity of the car first.

Given $\rho_A = 100m$

$$\therefore \rho_G = \rho_A - 0.6 = 99.4 \frac{m}{s};$$

Now note that the car is moving with constant speed. Hence $\frac{d\vartheta}{dt} = 0$.

Using (C), we may write an equation for acceleration of G as

$$\overrightarrow{a_G} = \frac{d\vartheta}{dt}\overrightarrow{e_t} + \frac{\vartheta^2}{\rho}\overrightarrow{e_n}$$

$$\text{OR, } \overrightarrow{a_G} = \frac{\vartheta^2}{\rho}\overrightarrow{e_n} \text{ since constant speed} \Rightarrow \frac{d\vartheta}{dt} = 0$$

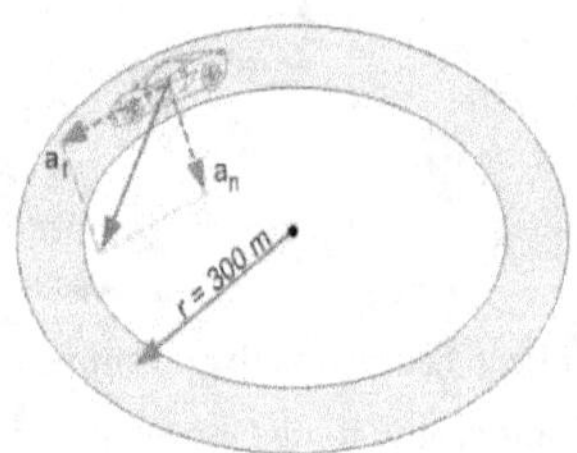

$$a_G = |\overrightarrow{a_G}| = 0.5 * 9.81 = \frac{\vartheta^2}{\rho_G}$$

$$\Rightarrow \vartheta^2 = \rho_G * 0.5 * 9.81 = 99.4 * 0.5 * 9.81$$

$$\Rightarrow \vartheta = 22.1 \text{ m/s}$$

Example: Car on Circular Track

A car travels around a horizontal circular path has a radius of 300m. If the car increases its speed at a constant rate of 7 m/s^2, starting from rest determine the time needed for it to reach a total acceleration of 8 m/s^2. What is the speed when it just attains 8 m/s^2? What is the distance travelled?

<u>Solution</u>: Since $a_t = 7$ m/s^2 = Constant, $\vartheta = at = 7t$. Thus $a_n = \frac{(7t)^2}{300} = 0.163t^2$ m/s^2 . Total acceleration, $a_c = 8$ m/s$^2 = \sqrt{a_t^2 + a_n^2} = \sqrt{7^2 + (0.163t^2)^2}$. So $t = \sqrt{\frac{\sqrt{8^2-7^2}}{0.163}} = 4.875$ sec.

Thus, $\vartheta = 7t = 34.1$ m/s

Also, $s = s_o + \vartheta_o t + \frac{1}{2}a_t t^2 = \frac{1}{2}*7*4.87^2 = 83$ m. So during this process of acceleration, the car traverses distance of $s = 83\ m$.

1.18.   Kinetics of a Particle: Three forms of Newton's Second Law

In the "Kinetics" part of analysis, we study the movements of bodies and the force causing these movements.

In addition to the information of acceleration and the components of accelerations, we use Newtons' Second Law extensively in Kinetics.

Very typically, the analysis consists of (i) Drawing FBD (ii) Writing equations of motion to find the accelerations, and (iii) using Kinematics analysis to find velocities and positions. In applying Newton's Second Law, which may be written in three forms[30]:

F – a Form

The Second Law stated as is: $\vec{F} = \frac{d}{dt}(m\vec{\vartheta}) = m\vec{a}$

It is assumed here that mass m = Constant and SI units are used.

Work – Energy Form

This form finds its use in Kinetic problems involving velocity, force and displacement. The work – Energy form can be derived from F – ma form. To do this, consider

$$F = ma = \frac{md\vartheta}{dt} = \frac{md\vartheta}{ds}\cdot\frac{ds}{dt} = m\vartheta\frac{d\vartheta}{ds}$$

$$\Rightarrow F.ds = m\vartheta d\vartheta$$

---

[30] Notice the number three. Wherever we go we see one of the three things: space, mass, and time. Each one is in turn in three forms: mass in solid, liquid, or gaseous form; space in length, breadth, or depth; time in past, present, and future. We have three laws of thermodynamics, three laws of mechanics. We have Trinity and Trimurthulu !

$$\Rightarrow \quad \overbrace{\int_{s_1}^{s_2} F\,(ds)}^{\text{Work done during the event}} \quad = \quad \overbrace{m \int_{\vartheta_1}^{\vartheta_2} \vartheta\, d\vartheta}^{\text{Change in kinetic energy}} \quad \Rightarrow W_{1-2} = \frac{1}{2}m(\vartheta_2^2 - \vartheta_1^2)$$

$$\Rightarrow \quad \overbrace{\frac{1}{2}m\vartheta_1^2}^{\text{Initial kinetic energy}} \quad + \quad \overbrace{W_{1-2}}^{\text{Work done during the event}} \quad = \quad \overbrace{\frac{1}{2}m\vartheta_2^2}^{\text{Final kinetic energy}}$$

Impulse – Momentum Form

This form is used in addressing Kinetic problem involving velocity, force, and time. This form again can be derived from F – ma form.

$$F = ma = m\frac{dv}{dt} \Rightarrow \int_{t_1}^{t_2} F\,(dt) = m \int_{\vartheta_1}^{\vartheta_2} d\vartheta$$

$$\Rightarrow \quad \overbrace{m\vartheta_1}^{\text{Initial Momentum}} \quad + \quad \overbrace{\int_{t_1}^{t_2} F\,(dt)}^{\text{Impulse during the event}} \quad = \quad \overbrace{m\vartheta_2}^{\text{Final Momentum}}$$

We begin with F – ma form of Newton's Second Law.

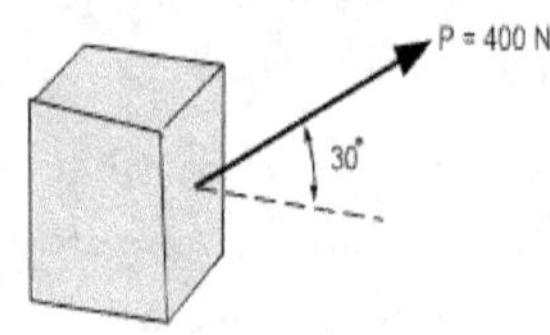

Example: Constant Force Acting on a Body

The 50 kg create shown in Figure beside resets on a horizontal plane for which coefficient of kinetic friction is $M_k = 0.3$.

If the crate is subjected to a 400W towing force as shown, determine velocity of crate in 3 & starting from rest.

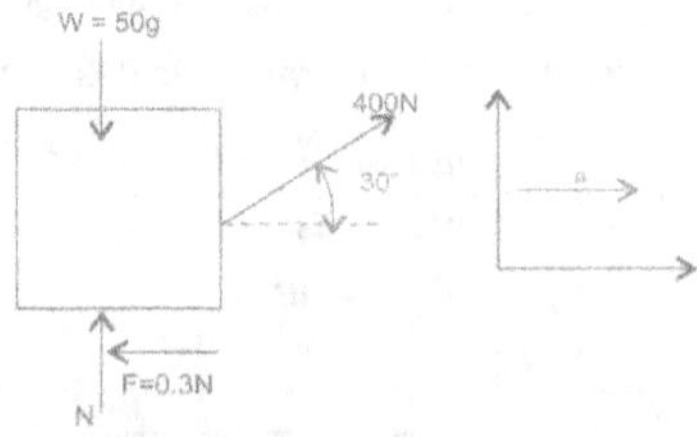

We draw the FBD of the crate and notice that the acceleration of the crate is along x-axis. We now can write equations of motion long x, y axes:

Along x – axis: $\sum F_x = ma_x \Rightarrow 400\, cos30° - 0.3N_c = 50a$ -----------(1)

Along y – axis: $\sum F_y = ma_y \Rightarrow N_c - 400.5 + 400 \sin 36° = 0 \Rightarrow N_c = 290.5N$

[note a_y=0]

Substituting the value of N_c into equation (1), we get $a = 5.19$m/s²

Since acceleration = Constant, $\vartheta = \vartheta_o + at = 0 + 5.19(3)$

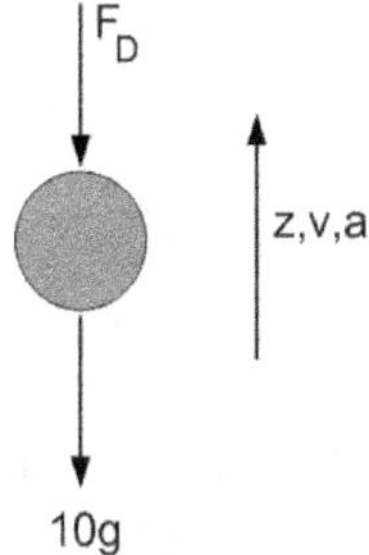

Or $\vartheta = 15.6$ m/s

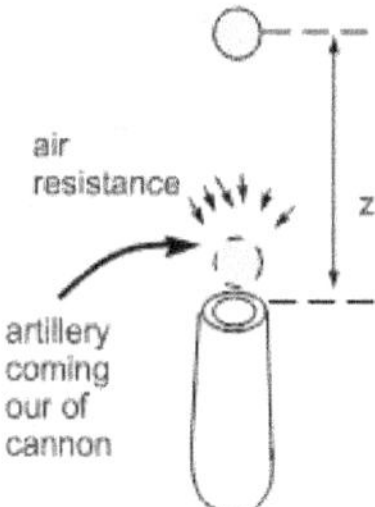

Example: Movement under Atmospheric Resistance

A 10 kg projectile is fired vertically upward from the ground with an initial velocity of 50 m/s. determine the maximum height to which it will travel if the atmospheric resistance is measured as $F_o = (0.01\vartheta^2)N$, where ϑ is the speed at any instant measured in m/s.

Solution: Notice that $z,\ \vartheta, a$ are taken positive upward. FBD of the projectile is shown in Figure beside. From the FBD, we write,

$$-F_D - 10g = 10a \Rightarrow a = -0.001v^2 - 9.81$$

Further, since the acceleration is not constant, we use the equation of the form: $a(ds) = v(dv)$ to write

$$a(dz) = \vartheta(d\vartheta) \Rightarrow (-0.001\vartheta^2 - 9.81)dz = \vartheta(d\vartheta)$$

$$\Rightarrow \int_0^h dz = -\int_{50}^o \frac{\vartheta d\vartheta}{0.001\vartheta^2 + 9.81} \Rightarrow h = 114m$$

1.19. Kinetics of a Particle: Work and Energy

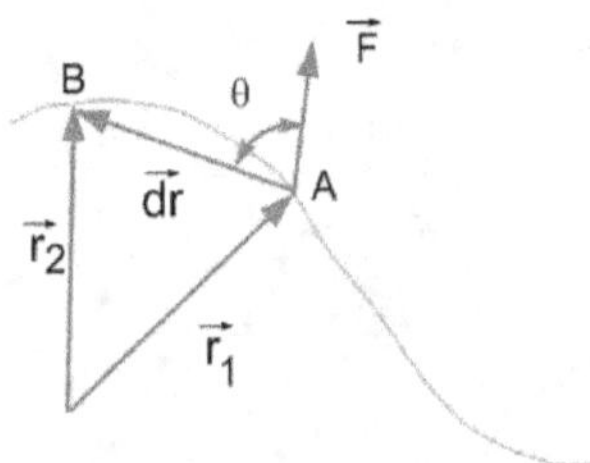

Work energy principle can be applied to solve problems that involve force, velocity, and displacement conveniently. We begin this topic by presenting the definition of work and a few instances of work done e.g., by weight g an object, by a spring, and a variable force.

Work

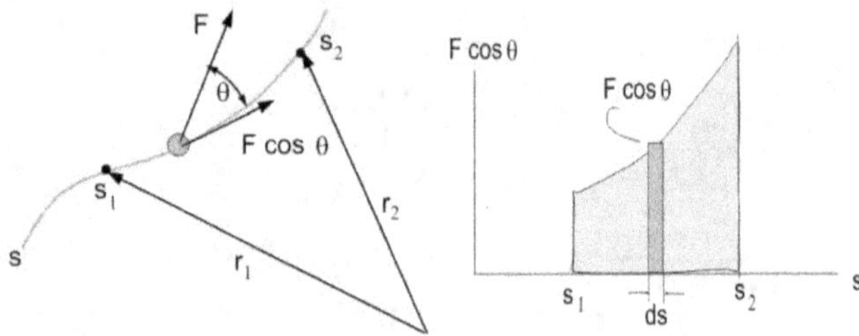

Work is said to be done <u>by</u> a force $\vec{F}$ <u>on</u> a particle only when the particle undergoes a displace in the direction of the force.

If the force, $\vec{F}$ in the figure moves a particle from a position A to a position B as shown in figure, work done by the force on the particle during this period is given by the dot product :
$dU = \vec{F}.\overrightarrow{dr}$ is the displacement of the particle as it moves from A to B.

The dot product is evaluated as: $dU = F(ds)\cos\theta$ where $F = |\vec{F}|$ and ds is the magnitude of displacement $|\overrightarrow{dr}| = \vec{r}_2 - \vec{r}_1$.

Note that the chord length from A to B is approximated by the arc-length from A to B.

Such approximations are justified when displacement considered is infitesimal. We will discuss on how to compute work done and give examples.

Work of a Variable Force

If a particle undergoes a finite displacement along its path from $\vec{r_1}$ to $\vec{r_2}$ (or s_1 to s_2) the work done is determined by integration.

If $\vec{F}$ is expressed as a function of position $F = F(s)$,

$$U_{1-2} = \{work\ done\ by\ the\ force\ on\ the\ pratice\ during\ r_1\ to\ r_2\} = \int_{\vec{r_1}}^{\vec{r_1}} \vec{F}.\vec{dr} = \int_{s_1}^{s_2} F \cos\theta (ds)$$

Geometric interpretation of this equation may be shown as the area under curve from s_1 to s_2. Notice the U_{1-2} can be negative depending on the angle θ.

Work of a Constant Force Moving Along a Straight Line

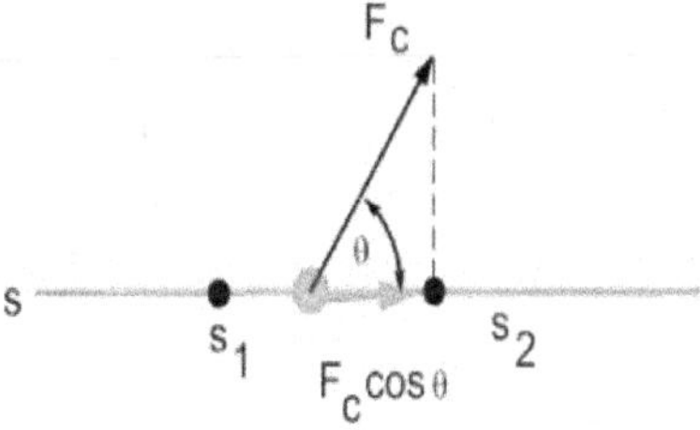

If a constant force, making a constant angle to the line of movement, moves a particle from s_1 to s_2, $U_{1-2} = \int_{s_1}^{s_2} F \cos\theta (ds) = F \cos\theta \int_{s_1}^{s_2} ds = F \cos\theta (ds)(s_2 - s_1)$

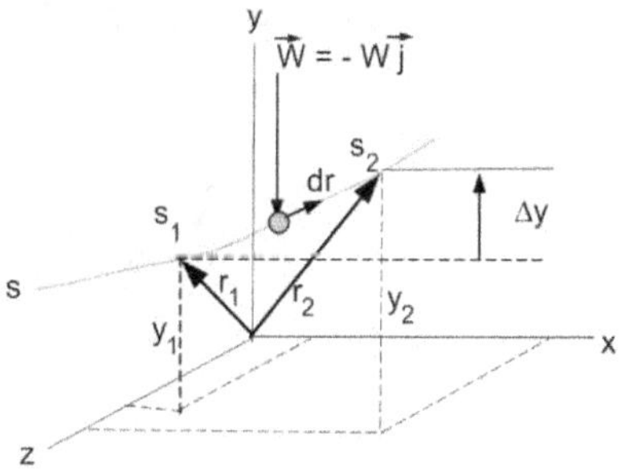

Work of a Weight

If the particle moves along the curve from point s_1 to s_2 , and intermediate point, the displacement nt is given by $\overrightarrow{dr} = (dx)\hat{\imath} + (dy)\hat{\jmath} + (dz)\hat{k}$,

work done is $U_{1-2} = \int_{\overrightarrow{r_1}}^{\overrightarrow{r_2}} \vec{F} . \overrightarrow{dr} = \int_{\overrightarrow{r_1}}^{\overrightarrow{r_2}} (-w\hat{\jmath}) . \left[(dx)\hat{\imath} + (dy)\hat{\jmath} + (dz)\hat{k} \right]$

$= \int_{y_1}^{\overline{y_2}} -w(dy) = -w(y_2 - y_1) = -w(\Delta y)$

Note: for the case shown in figure, work done cane out to be negative because W is downward and displacement is upward. K is the spring constant = force needed for unit deformation of spring.

Work of a Spring Force

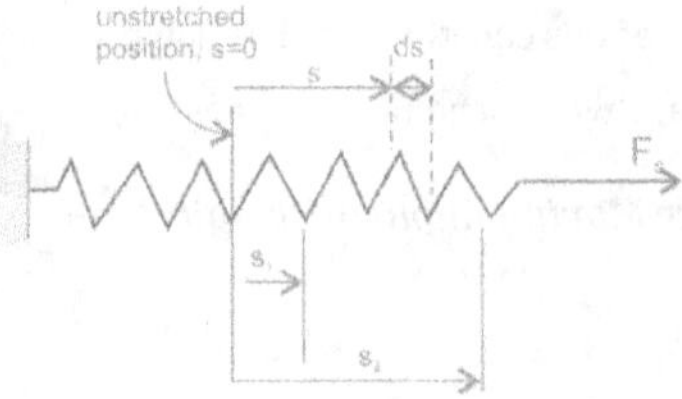

If a spring is stretched by a distance of s from its unstretched position, force required is given by F_s = ks. To compute the work done <u>on</u> the spring <u>by</u> the force F_s in stretching the spring from a position s_1 to s_2 may be written as: $U_{1-2} = \int_{s_1}^{\overrightarrow{s_2}} F(ds) = \int_{s_1}^{\overrightarrow{s_2}} ks(ds) = \frac{1}{2}ks_2^2 - \frac{1}{2}ks_1^2$ (remember this).

NOTE that work done by the force on the spring either in elongating or compressing the spring is positive. (why?)

If a body is attached to the spring and the spring under goes deformation from s_1 to s_2 then, work done by the spring on the body is obtained as $U_{1-2} = -\left(\frac{1}{2}ks_2^2 - \frac{1}{2}ks_1^2 \right)$ (remember this).

Example: Work Done in Lifting an Object

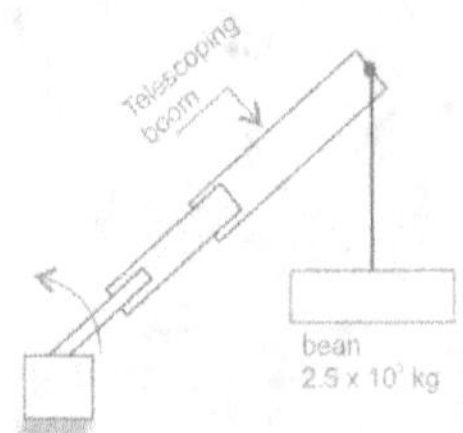

For a short time, the crane in figure lifts a 2.5mg beam with a force of $F(28 + 3s^2)KN$. Determine the speed of the beam when it has risen $s = 3m$. Also, how much time does it take to attain this height?

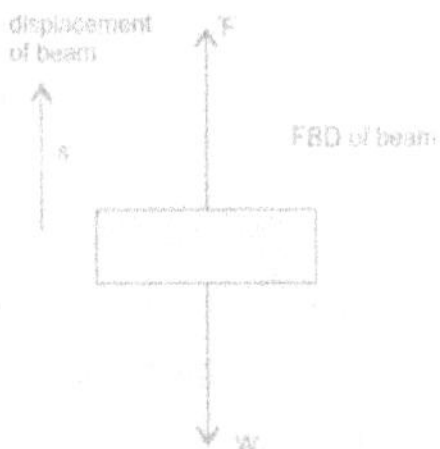

Draw FBD of beam. Since F and s are in same direction, work done by F on beam is positive. However, work done by W is negative because W and s are in opposite directions. We now apply work energy principle to the beam.

As the beam is raised by the crane from an initial position of $s_1 = 0, v_1 = 0$ to a position of $s_2 = s_m$ and $\vartheta_2 = \vartheta$ m/s we may write:

$$\underbrace{\left\{\begin{matrix}\text{Initial KE} \\ \text{of the beam}\end{matrix}\right\}}_{\text{At rest.So zero}} + \underbrace{\left\{\begin{matrix}\text{Work done} \\ \text{by F on beam}\end{matrix}\right\}}_{\text{Positive } (W_1)} + \underbrace{\left\{\begin{matrix}\text{Work done} \\ \text{by W on beam}\end{matrix}\right\}}_{\text{Negative } (W_2)} = \underbrace{\left\{\begin{matrix}\text{Final KE} \\ \text{of beam}\end{matrix}\right\}}_{\frac{1}{2}m\vartheta^2}$$

$W_1 = \int_{s_1}^{s_2} F(ds) = \int_0^s (28 + 3s^2)10^3 \, ds = 28(10^3)s + (10^3)s^3$

And $W_2 = -2.5 * 10^3 * 9.81s$

$\therefore (28s + s^3)10^3 - 24.525(10^3 s) = 1.25\vartheta^2(10^3)$

$\therefore \vartheta = \sqrt{(2.78s + 0.8s^3)}$

When $s = 3, \vartheta = 5.47 \frac{m}{s}$

Since, $\vartheta = \frac{ds}{dt}$, we have,

$\int_0^t dt = \int_0^3 \frac{ds}{\sqrt{2.78s + 0.8s^3}} \Rightarrow t = 1.78s$

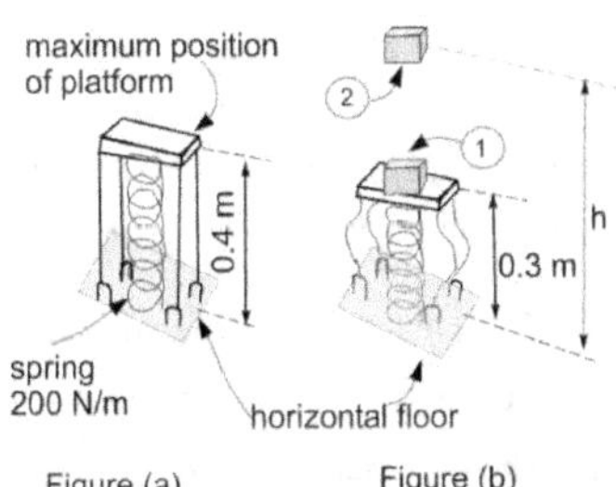

Example: Work done by Spring Force

The platform P shown in figure has negligible mass and is tied down so the 0.4m long cords keep a 1-m long spring compressed 0.6 m when nothing is there on the platform. If a 2 kg block is placed on the platform and released from rest after the platform is pushed down 0.1 m, determine the maximum height h the block rises in the air, measured from ground.

Solution

We use work energy principle for the block as it moves from position (1) in figure 6 to position (2). At position (1), compression of spring, $s_1 = 1 - 0.3 = 0.7m$. though the block reaches height h as shown in figure (b) the platform is in contact with block only up to maximum position shown in figure (a). Hence compression of spring when block reaches position (2) is $s_1 = (1 - 0.4) = 0.6m$. hence work done by the spring on the block, W_1, is given by

$$W_1 = -\left(\frac{1}{2}ks_2^2 - \frac{1}{2}ks_1^2\right) = -\frac{1}{2}*200(0.6^2 - 0.7^2) = 13N - M$$

$$W_2 = -W(h - 0.3) = -2*9.81(h - 0.3) = (-19.62h + 5.856)$$

Notice that when the block reaches the maximum height, it will be at rest momentarily before it begins to fall down. Therefore, Initial energy = final energy = 0

$\therefore$ Work Energy principle applied to block for positions 1 to 2 may be written as:

$$\left\{\begin{array}{c}\text{Initial KE of}\\\text{of block}\\\text{at position 1}\end{array}\right\} + \left\{\begin{array}{c}\text{Work done by}\\\text{spring on block}\\\text{as it moves from 1 to2}\end{array}\right\} + \left\{\begin{array}{c}\text{Work done by the}\\\text{weight of block on}\\\text{the block as the block}\\\text{moves from 1 to 2}\end{array}\right\} = \left\{\begin{array}{c}\text{Final KE}\\\text{of block at}\\\text{position 2}\end{array}\right\}$$

zero because starting from rest :velocity =0   zero because when the block reac hes maximum position it is momentarily at rest

$$\Rightarrow 0 + W_1 + W_2 = 0 \Rightarrow 13 - 19.62h + 5.886 = 0$$

$$\therefore h = \frac{18.886}{19.62} = 0.962\ m$$

1.20. Planar Rigid Body Motion

So far, we studied about movement of a "particle". To study gears, cams, and several other linkages, we need to consider kinetics of a rigid body.

Furthermore, once a kinetics of a rigid body is thoroughly understood, dynamics of linkages may be studied using appropriate equation of motion to relate the forces acting on the body and the resulting movement of the body planar kinetic study is a beginning point.

Planar Motion

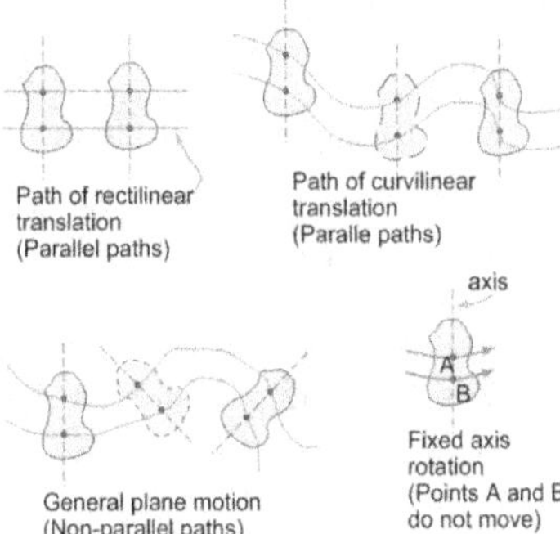

Planar motion is said to occur when all the particles of a rigid body move along path which are equidistant from a fixed plane.

There are three types of planar motion: (i) Translation (ii) Rotation about fixed axis, and (iii) General plane motion.

Translation

Translation is said to occur if, during motion, every line segment on the body remains parallel to its original direction. This may further be classified into <u>rectilinear translation</u> where in path of motion of any two particles on the body are along equidistant straight lines and <u>curvilinear translation</u> where in the path of motion are along curved lines.

Fixed Axis Rotation

When a rigid body rotates about a fixed axis, all the particles of the body except those that lie on the axis of rotation, move along circular paths. These three types of planar motion are shown in figure beside, we briefly look into translation and fixed axis rotation in subsequent sections.

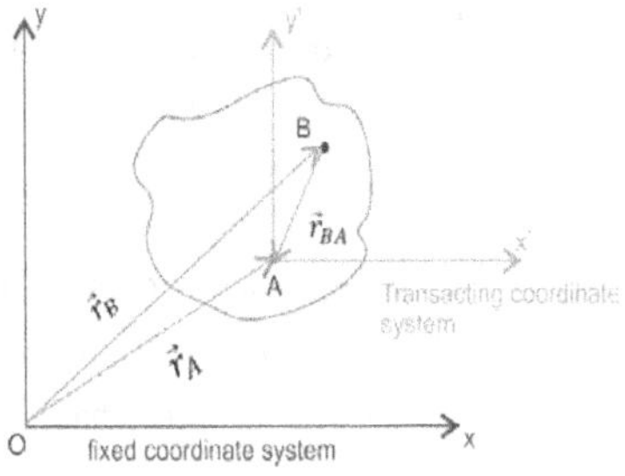

93

Translation

Consider about (shown in pink) either in rectilinear or curvilinear translation as shown in figure beside this body is undergoing motion relative to a Fixed Coordinate system $x - y$ with origin at O. consider two points A and B on the body with position vectors $\vec{r_A}$ and $\vec{r_B}$. The vector $\vec{r_{BA}}$, going from A to B is called relative position vector. The origin of the moving coordinate system, point A, is termed the base point and we write:

$$\vec{r_B} = \vec{r_A} + \vec{r_{BA}}$$

This equation relates the position of any point B with respect of that of A. we can differentiate this equation to find velocities:

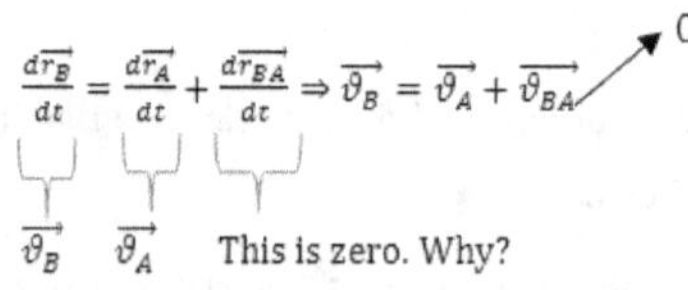

$$\frac{d\vec{r_B}}{dt} = \frac{d\vec{r_A}}{dt} + \frac{d\vec{r_{BA}}}{dt} \Rightarrow \vec{\vartheta_B} = \vec{\vartheta_A} + \vec{\vartheta_{BA}}$$

$$\vec{\vartheta_B} \qquad \vec{\vartheta_A} \qquad \text{This is zero. Why?}$$

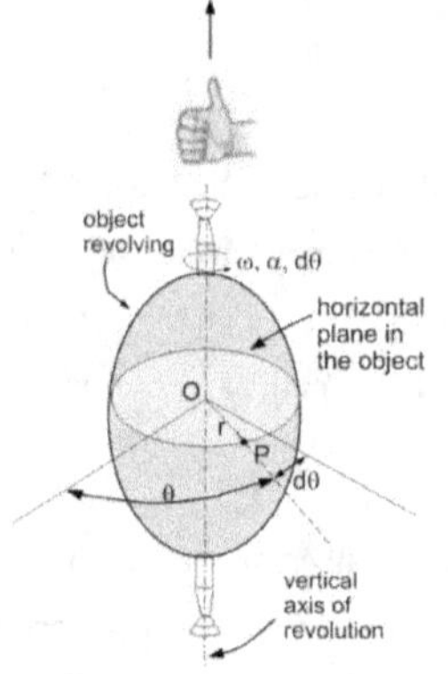

Similarly, we may see that $\vec{a_B} = \vec{a_A}$. Hence, we may conclude that all particles / points in a rigid body undergoing either rectilinear translation or curvilinear translation move with same velocity and acceleration. Thus, kinematics of particle motion, hither to discussed, may also be used to specify the kinematics of points in a translating body.

Fixed Axis Rotation

A point is a Euclidean object which has zero dimension so it has no angular motion only lines, planes or solids undergo angular motion. As an example, consider the body shown beside. Which is under going rotation. The rotation, θ, of this body may be shown as rotation of radial line r. The angular displacement $\vec{d\theta}$, angular velocity $\vec{\omega}$, and angular acceleration $\vec{\alpha}$ are vector quantities and their direction is obtained by the direction of thumb of right hand shown in the figure beside. We may write kinematic equations.

$\omega = \dfrac{d\theta}{dt}$; $\alpha = \dfrac{d\omega}{dt} = \dfrac{d^2\theta}{dt^2}$; $\alpha(d\theta) = \omega(d\omega)$ analogous to translation. Also, when angular acceleration is constant, we may write the equations.

$\omega = \omega_o + \alpha_c t$

$\theta = \theta_o + \omega_o t + \dfrac{1}{2}\alpha_c t^2$

$\omega^2 = \omega_o^2 + 2\alpha_c(\theta - \theta_o)$

$\alpha_c =$ Constant Angular Acceleration

Carefully notice the directions of vector quantities θ, ω and α. When the figures curl in the direction of rotation, the direction of thumb gives the direction of θ, ω, α. This also indicates why anticlockwise direction is taken as positive:

In the fingers of right hand are curled in clockwise direction, it will have a serious negative effect on us!

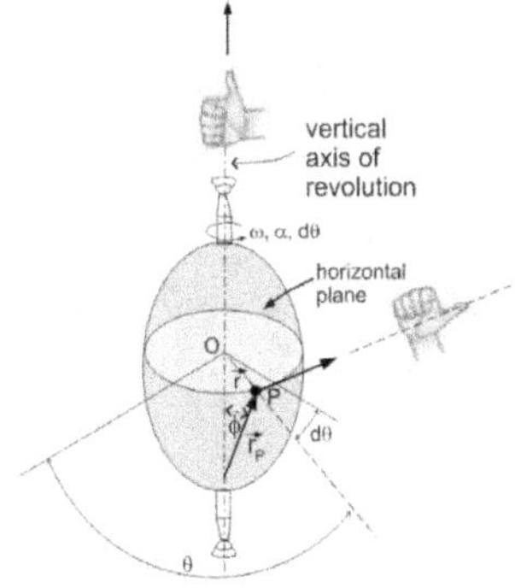

To study the motion of a point P on the body as shown in figure beside. In this figure, the vector $\vec{r}$ is in the shaded horizontal plane, $\vec{r_p}$ is a vector going from any point on the axis to the point P. We know that $\vartheta = |\vec{\vartheta}| = \omega r$ where $\omega = |\vec{\omega}|$ and $r = |\vec{r}|$. In vector rotation, we write:

$$\frac{d\vec{r_p}}{dt} = \vec{\vartheta_p} = \vec{\omega} \times \vec{r_p} \qquad (1)$$

The direction of $\vec{\vartheta_p}$ is obtained by the rule of cross – multiplication of vectors: When the fingers are curled from $\vec{\omega}$ to $\vec{r_p}$, thumb gives direction of $\vec{\omega} \times \vec{r_p}$ as shown by hand 2 in the figure. Differentiating 1, we obtain total acceleration of point P as

$$\overrightarrow{a_p} = \frac{d\overrightarrow{\vartheta_p}}{dt} = \frac{d\overrightarrow{\omega}}{dt} \times \overrightarrow{r_p} + \overrightarrow{\omega} \times \frac{d\overrightarrow{r_p}}{dt}$$

$$\text{Or, } \vec{a}_p = \overbrace{\vec{\alpha} \times \vec{r_p}}^{\vec{a}_t} + \overbrace{\vec{\omega} \times (\vec{\omega} \times \vec{r_p})}^{\vec{a}_n}.$$

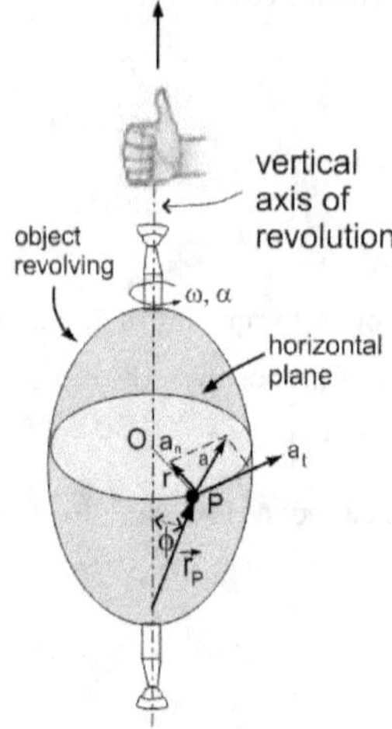

Referring to the figure to the left, we see that

$$\overrightarrow{a_p} = \overrightarrow{a_t} + \overrightarrow{a_n} = \vec{\alpha} \times \vec{r} - \omega^2 \vec{r} .$$

And $a_p = |\overrightarrow{a_p}| = \sqrt{a_t^2 + a_n^2}$. Notice that $\mathbf{a_t}$ and $\mathbf{a_n}$ are in the horizontal plane. $\mathbf{a_n}$ is a vector going from P towards O and $\mathbf{a_t}$ is perpendicular to OP.

Example: Anemometer

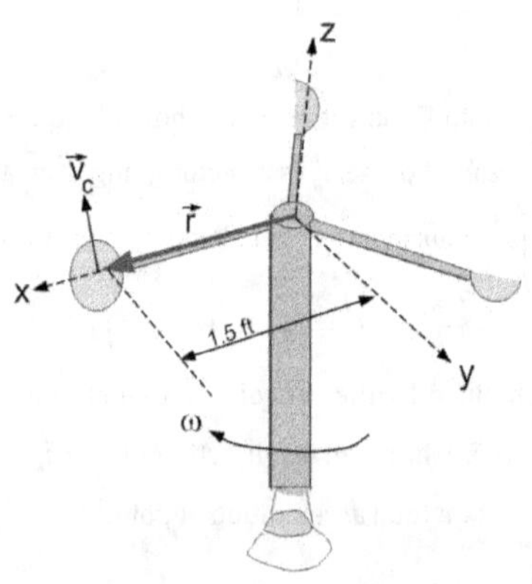

The anemometer, shown in figure beside, measures the speed of the wind due to the rotation of the three cups. If during a 3-second time speed a wind guest causes the cups to have an angular velocity given by $\omega = (2t^2 + 3)$ rod/sec, where t is in seconds, determine (a) the speed of the cups when t = 2, (b) the total distance travelled by each cup during the 3 – second time period, and (c) the angular acceleration of the cups when t = 2 seconds. Neglect the size of the cups for calculation.

<u>Solution:</u> velocity of the cup $\overrightarrow{\vartheta_c}$ may be written as:

$$\overrightarrow{\vartheta_c} = \vec{\omega} \times \vec{r}$$

(a) With the choice of coordinate axis, $\vec{\omega} = -(2t^2 + 3)\vec{k}$ rod/sec and $\vec{r} = 1.5\vec{i}$ ft. Therefore, $\overrightarrow{\vartheta_c} = -(2t^2 + 3)\vec{k} * 1.5\vec{i} = -1.5(2t^2 + 3)\vec{j}$. So, at $t = 2, \overrightarrow{\vartheta_c} = -1.5\vec{j}$ ft/sec. The direction of $\overrightarrow{\vartheta_c}$ is shown in figure above. [convince yourself that the direction of $\overrightarrow{\vartheta_c}$ shown is obtained by rules of cross - product].

(b) $\omega = (2t^2 + 3)$ rod/sec $\Rightarrow d\omega = 4t(dt)$. Since this is NOT a case of constant acceleration we invoke the kinematic relation: $\alpha(d\theta) = \omega(d\theta)$ to get

$$(4t)(d\theta) = (2t^2 + 3) * 4t * (dt) \Rightarrow \theta = \left[\frac{2t^3}{3} + 3t\right]_o^3 = 27 \text{ rod For one revolution,}$$

distance travelled is $2\pi r$. Therefore, distance travelled during 27 radius of rotation is $\frac{27*1.5*2\pi}{2\pi} = 40.5$ ft

(c) $\alpha = \frac{d\omega}{dt} = 4t$ rod/sec². So, when $t = 2$ seconds, $\alpha = 4 * 2 = 8$ rod/sec².

1.21. Practice Problems

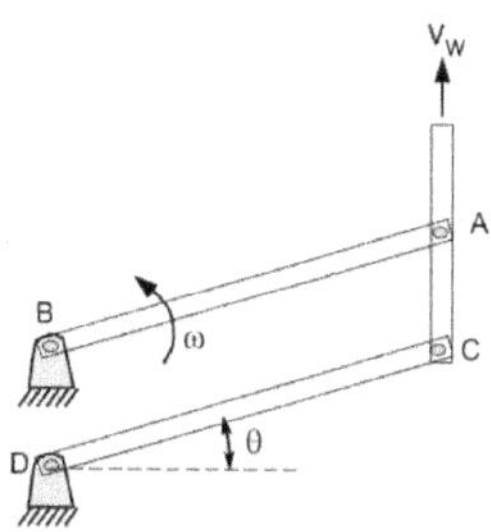

1) The linkage mechanism for a car window winder is shown in figure beside. Here, the lever AB rotates by the motion of a handle (not shown in figure) and raises the window. If, at the instant shown, the lever AB has an angular velocity of ω = 0.2 rad/sec, determine the speed of points A and C and the speed ϑ_w of the window. Take

$\theta = 30°$.Also take lengths of levers AB and CD as 200 mm. [ans: $\vartheta_w = 34.6 \; mm/s$; $\vartheta_A = \vartheta_E = 40mm/s$]

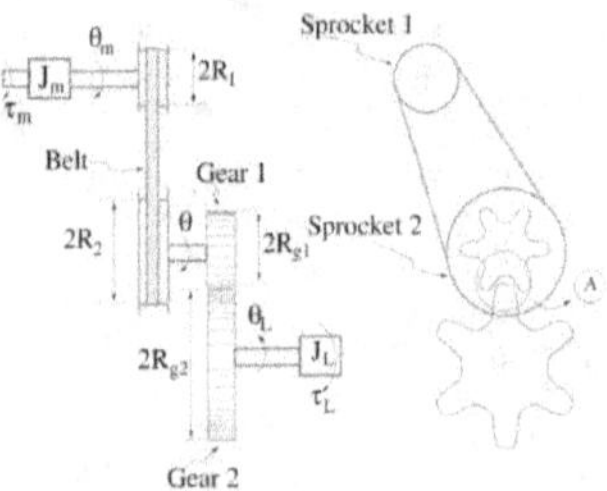

2) Due to increase in power, the gear 1 in the Figure rotates with an angular acceleration of $\alpha = 0.06\theta^2 \; \frac{rad}{sec^2}$ where θ is in radians. If the gear 1 is initially turning at $\omega_o = 50$ rad/s, determine the angular velocity of gear 2 after the gear 1, undergoes an angular displacement $\Delta\theta = 10 \; rev$. $R_1 = 24$ mm and $R_2 = 60$ mm. [Ans:22.2 rad/s]

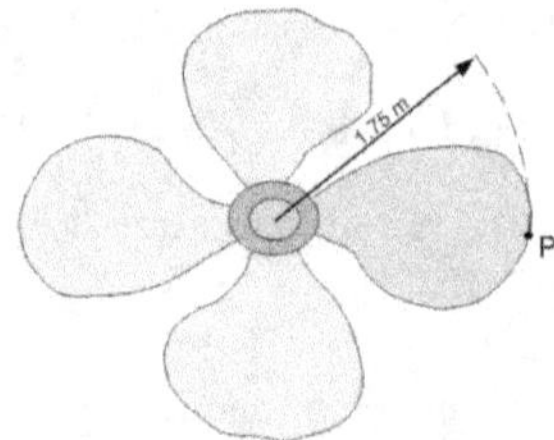

3) Just after the fan is turned on, the motor gives the blade an angular acceleration $\alpha = 20e^{-0.6t}$ rod/s², where t is in seconds. Determine the speed of the tip of one of the blades when t = 3 seconds. How many revolutions has the blade turned in 3 seconds? When t = 0, the blade is at rest. [Ans: 27.8 m/s; 8.5 rev]

1.22. Exercise Problems

1) The three forces shown in Figure keep a particle in equilibrium. Find the magnitudes of forces F_1 and F_2. [1039.5 N, 989.4 N]

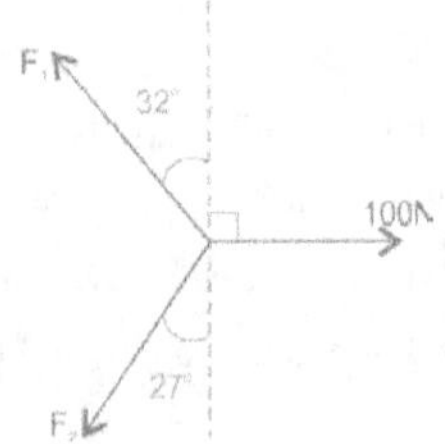

2) Four forces F_1, F_2, F_3 and F_4 keep a body in equilibrium. Find the magnitude and direction of force F2. [1619.9N; 45.2°]

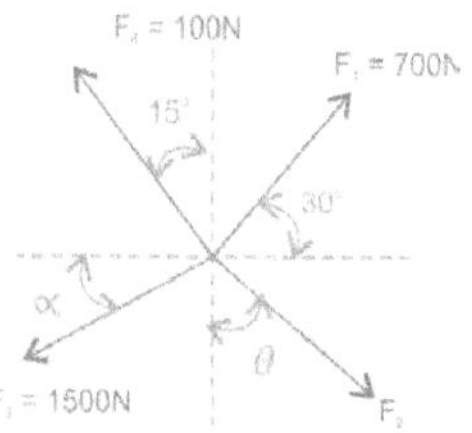

3) Find the resultant of the six forces shown in Figure below. [8500N; 280.4°]

Force	Magnitude	Angle (θ_i)
F_1	10,000 N	45°
F_2	9,000 N	23°
F_3	8,000 N	63°
F_4	7,000 N	70°
F_5	6,000 N	0°
F_6	5,000 N	270°

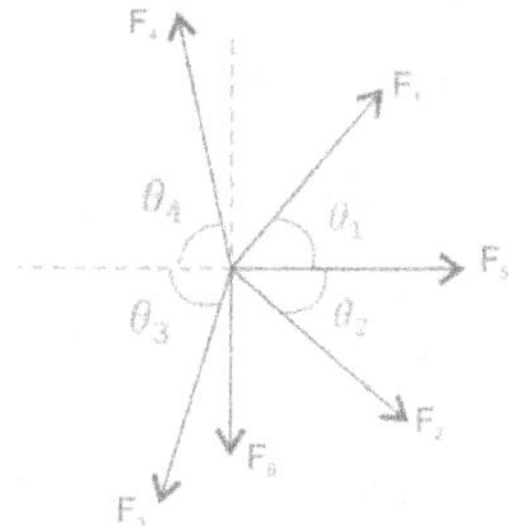

4) The three forces acting on the screw eye produce a resultant force of $F_R = 0$. If $F_2 = \frac{2}{3}F_1$ and F_1 is to be 90° from F_2 as shown, determine the required magnitude of F_3 expressed in terms of F_1 and the angle θ. [$F_3 = 1.2F_1$, $\theta = 63.87$]

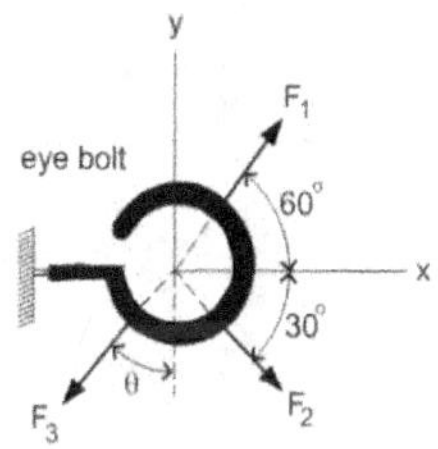

5) The three forces are applied to the bracket. Determine the range of values for the magnitude of force P so that the resultant of the three forces does not exceed 2400N. $[1.22N \leq P \leq 3.17N]$

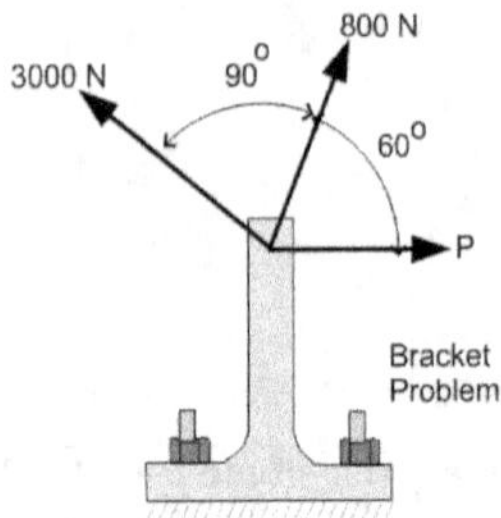

6) If the magnitude of the resultant force acting on the full-eye bolt is 600N and its direction measured clockwise from the positive x −axis is $\theta = 30°$, determine the magnitude of the force F_1 and the angle Ø. $[F_1 = 731N, Ø = 42.4°]$.

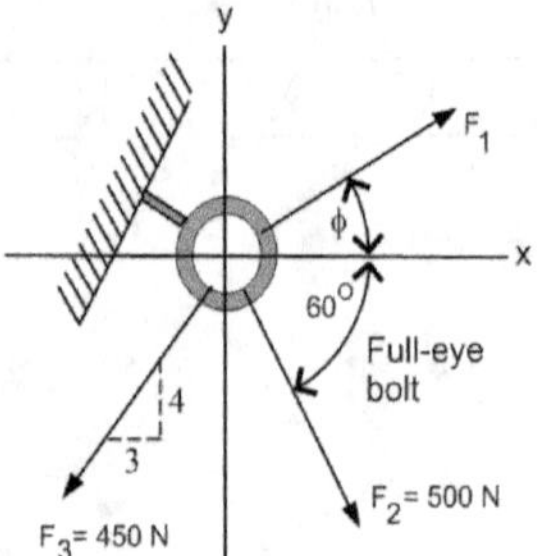

7) Determine the magnitude of F_1 and its direction θ so that the resultant force is directed vertically upwards and has a magnitude of 800N. [F₁ = 731N, θ = 29.1°]

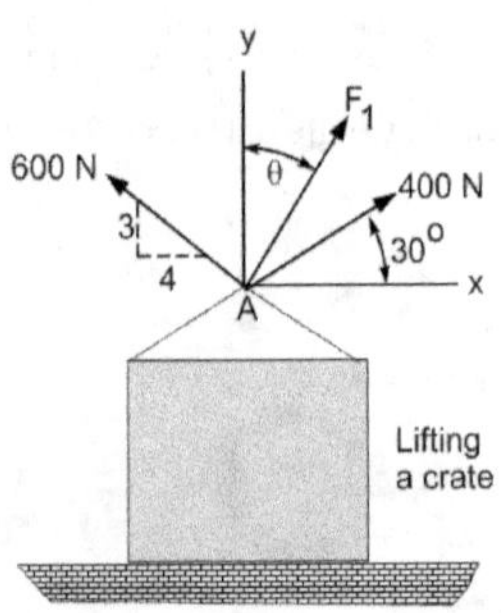

8) Three cylinders, A, B and C are piled in a rectangular ditch as shown in Figure beside.
   $W_A = 300N$, $r_A = 0.4m$, $W_B = 800N$, $r_B = 0.6m$, $W_C = 400N$, $r_C = 0.5m$.   neglecting
   friction, determine all contact forces.

   $[R_p = 323N; R_q = 508N; R_s = 1923N; R_t = 2000N; R_v = 1500N; R_w = 1600N]$

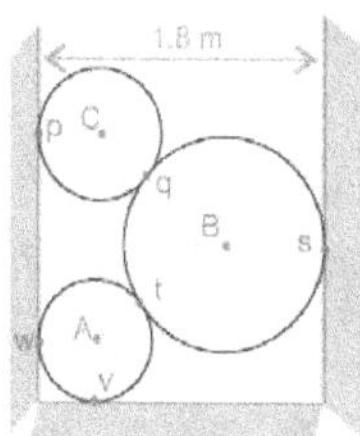

9) The forces on the gusset plate of a joint in a bridge truss act as shown in Figure beside.
   Find the values of P & F to maintain the equilibrium of the joint. [F = 800N; P = 3342N]

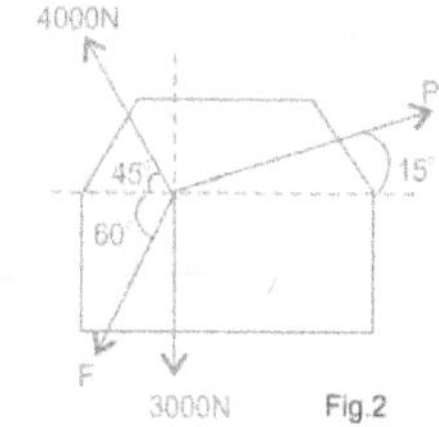

10) Find the reactions at ends of a simply supported beam loaded as shown neglect friction
   at both ends and neglect the weight of beam. $[R_A = \dfrac{4000}{3}N; \ R_B = \dfrac{5000}{3}N]$

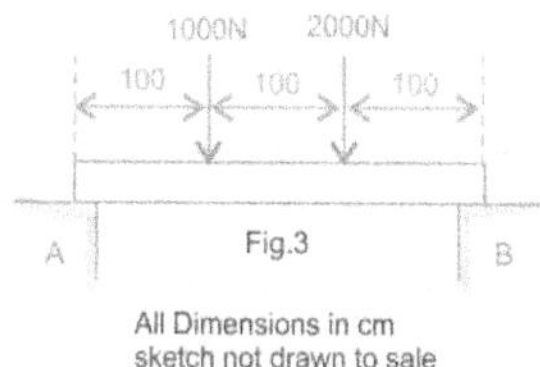

11) The 300N sphere in Figure 4 is supported by the pull P and a 200N forces as shown. If
   $\alpha = 30^\circ$, compute the values of P and θ. $[\theta = 38.26^\circ; P = 161.48N]$

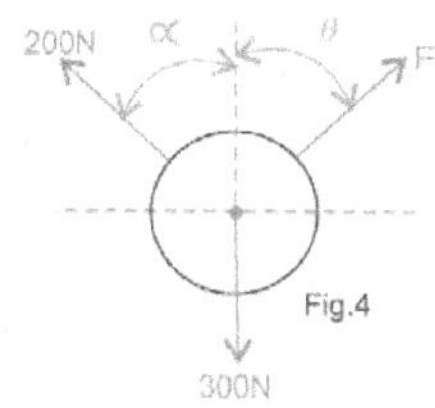

12) Three bars, AB, BC and CD are connected by pin joints to form a mechanism as shown in Figure 5. The mechanism is connected to the ground at A and D using pin joints. A force of 200N acts on the pin at B as shown in Figure 5. Determine the value of force P that must be applied on pin at C as shown in figure 5 to maintain equilibrium. [P = 305N]

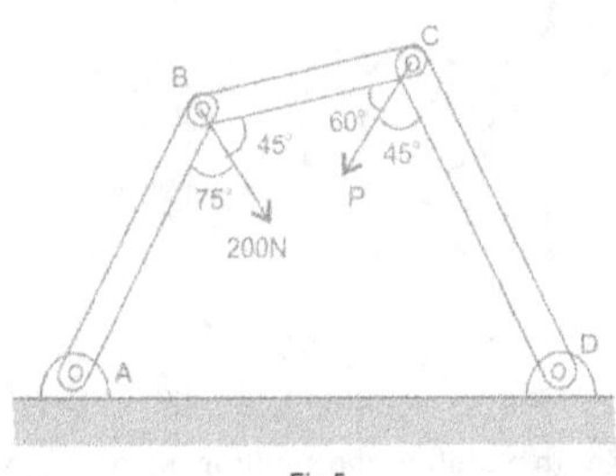

Fig.5

13) Determine the minimum moment produced by the force F about point A. specify the angle θ. Also find the maximum moment.

$$[M_{max} = 1.44 \ kN - m, \theta_{max} = 56.3°, M_{min} = 0, \theta_{min} = 14.6°]$$

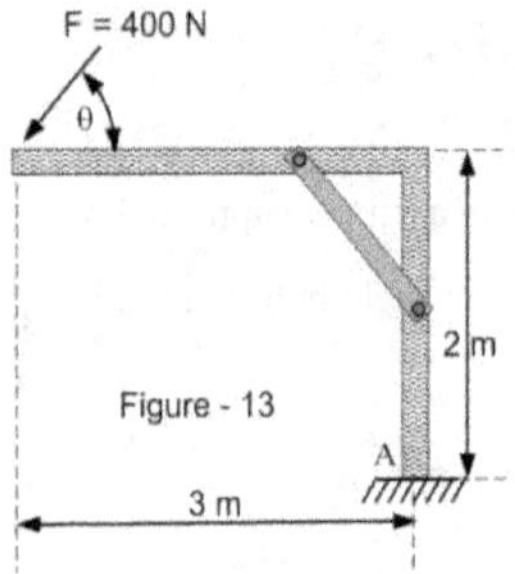

14) A force of 40N is applied to the wrench. Determine the moment of this force about point O. [- 7107.1 N-MM]

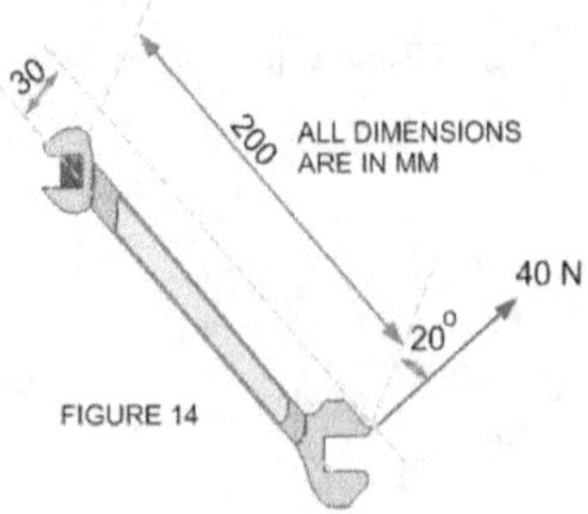

15) Determine the magnitude and directional sense of the resultant moment of the forces at A and B about point O and then about point P in Figure 3. [4.47KN-M, 3.31 KN-M]

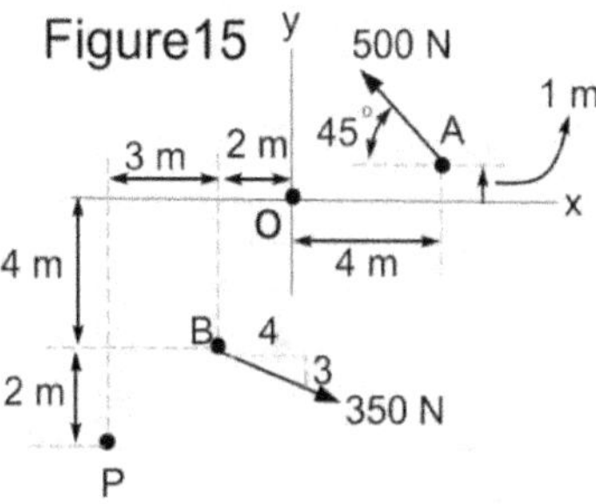

16) The articulating boom platform can support a weight of 550 N. If the boom is in the position shown, determine the moment of this force about points A, B and C. [-1650 N-m, -9270 N-m, 6450 N-m].

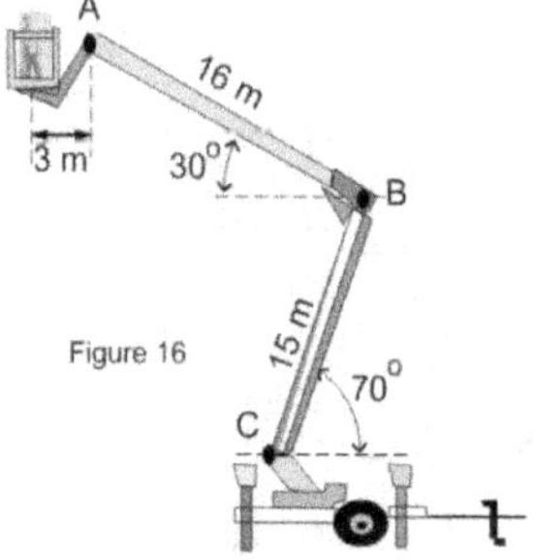

17) Determine the magnitude and direction of the smallest force P required to start the wheel in the figure beside over the block. What is the reaction at the block? [1896N at 71.5° with the horizontal; 638N]

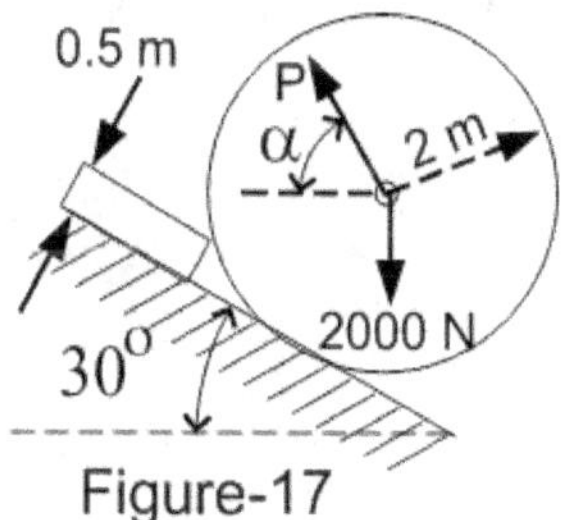

18) Two smooth balls, each of weight P and radius r are placed inside a cylinder open at both ends. The assembly rests on a horizontal surface as in figure besides. If the cylinder is of weight W and radius R < 2r, find the force exerted by either ball on the cylinder and the smallest value of W that will prevent the cylinder from tipping over.

$$\left[F = \frac{P(R-r)}{\sqrt{2rR-R^2}}; W = \frac{2P(R-r)}{R}\right]$$

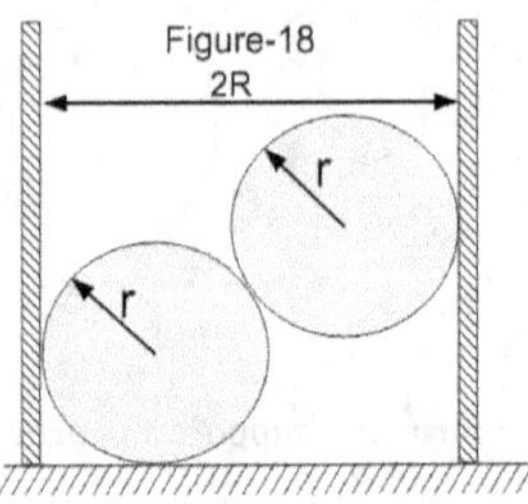

19) A 300N – box is held at rest on a smooth incline by a force P making an angle θ with the incline as shown in the figure beside. If θ is chosen to be 45°, determine the value of P.
[212N]

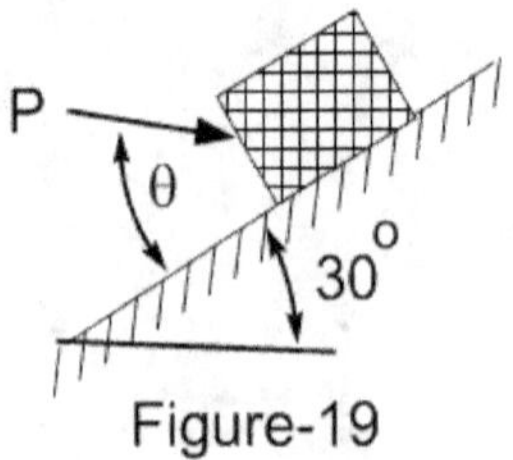

20) Two cylinders, A of weight 400N and B of weight 200N rest on smooth inclines. They are connected by a bar of negligible weight hinged to each cylinder at its geometric center by smooth pins. Find the force P acting as shown that will hold the system in given position [$F_{BAR} = 490N$; $R_A = 546N$; $R_B = 538N$; $P = 107N$]

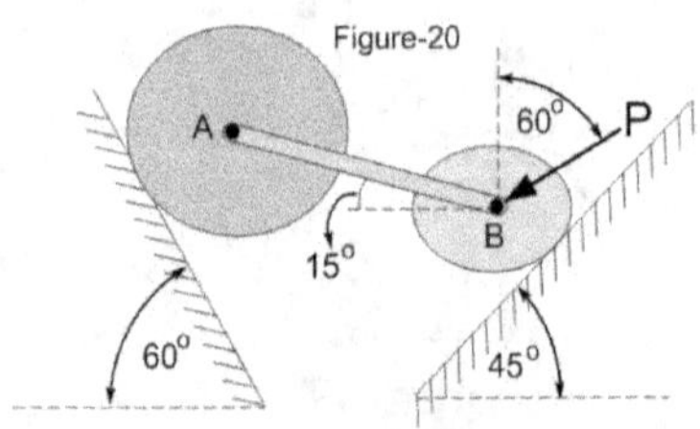

21) A 20-N horizontal force is applied perpendicular to the handle of the socket wrench. Determine the magnitude and direction of the moment created by this force about point O. $[4.27N - M; \alpha = 95.2°; \rho = 110°; \sigma = 20.6°]$

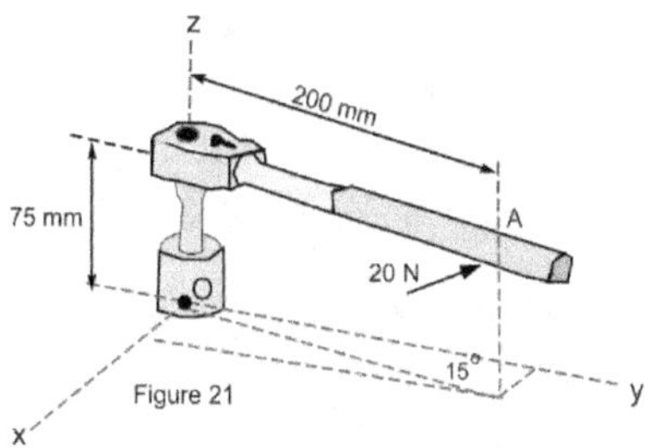

22) The force $\vec{F} = (6\vec{i} + 8\vec{j} + 10\vec{k})N$ creates a moment of $M_o = [-14\vec{i} + 8\vec{j} + 2\vec{k}]N - M$. If the force passes through a point having an x −coordinates of 1m, determine the y and z co-ordinates of the point. Also, realizing that $M_o = Fd$, determine the perpendicular distant "d" from point O to line of action of F. [y = 1m; z = 3m; d = 1.15m]

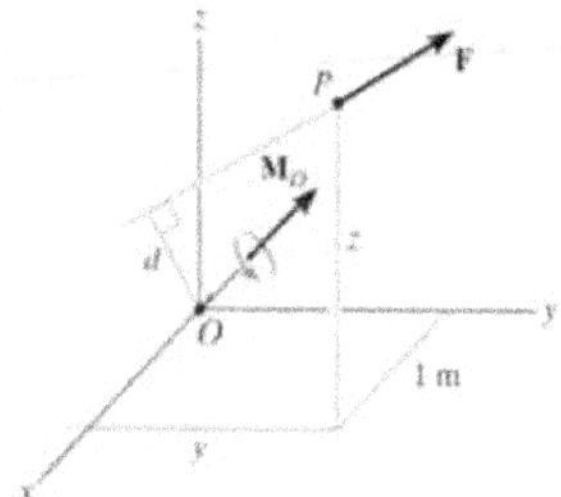

23) Determine the moment produced by each force about point O located on the drill bit. The point A is lying in xy − plane and the point B is lying on the yz − plane.

$$[\overrightarrow{M_A} = (-18\vec{i} + 9\vec{j} - 3\vec{k})N - M; \quad \overrightarrow{M_B} = (18\vec{i} + 7.5\vec{j} + 30\vec{k})N - M]$$

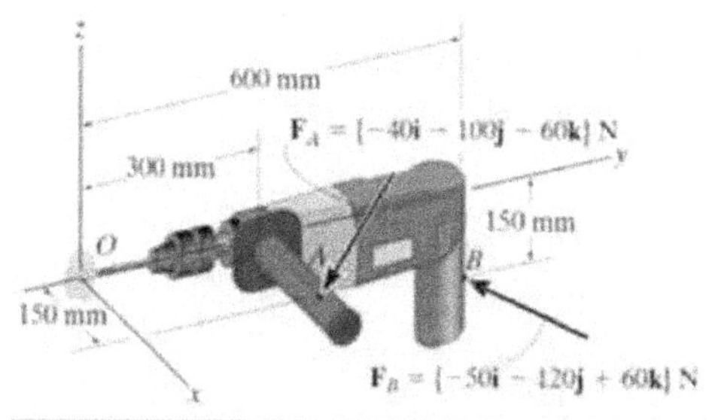

24) If P = 200N determine the friction developed between the 50kg crate and the ground. The coefficient of static friction between the crate and the ground is μ_s = 0.3. [160N]

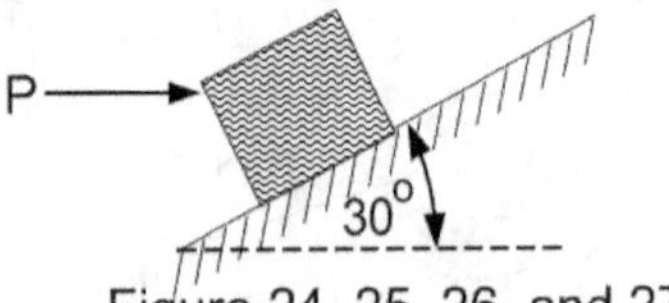

Figure 24, 25, 26, and 27

25) Determine the minimum horizontal force needed to hold the crate from sliding down the plane. The crate has a mass of 50kg and the static coefficient of friction between the crate and the plane is μ_s = 0.25. [140N]

26) Determine the force P to move the crate in figure up the plane. [474N]

27) A horizontal force of P = 100N is sufficient to hold the crate in figure from sliding down and a horizontal force of P = 350N is just sufficient to push the crate up the plane. Determine the coefficient of friction between the plane and the crate and the weight of the crate. [0.256]

28) Determine the minimum force P required to prevent the 30kg rod from sliding. The contad surface at B is smooth whereas the coefficient of static friction between the rod and the wall at A is U_s = 0.2. [155N]

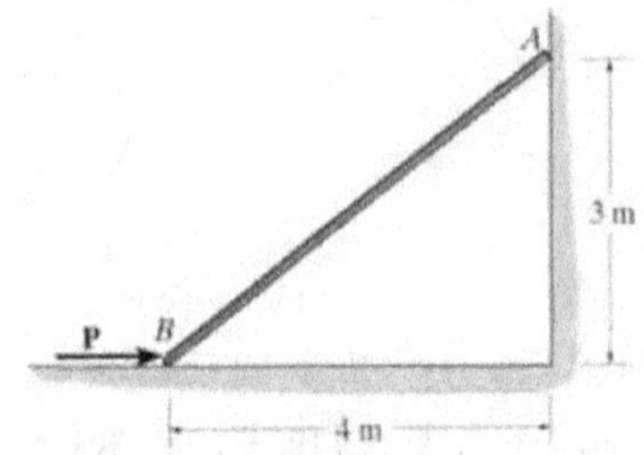

29) Determine the maximum force that can be applied without causing the blocks to move. The coefficient of static friction between each block and the ground is M = 0.25 and the mass of each crate is 50kg. [247N]

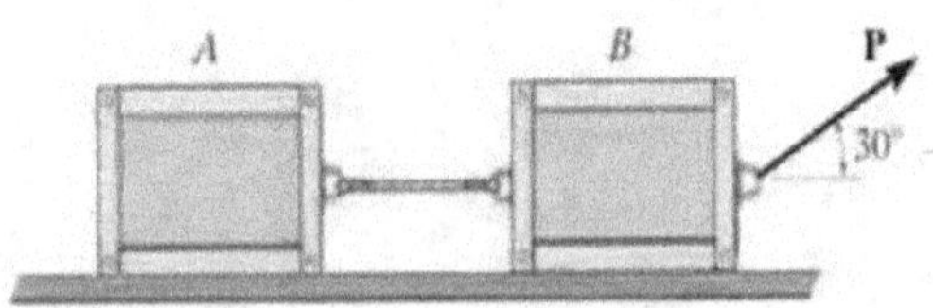

30) Determine the location of centroid of quarter circle

Ans: $\bar{x} = \dfrac{4r}{3\pi}$; $\bar{y} = \dfrac{4r}{3\pi}$

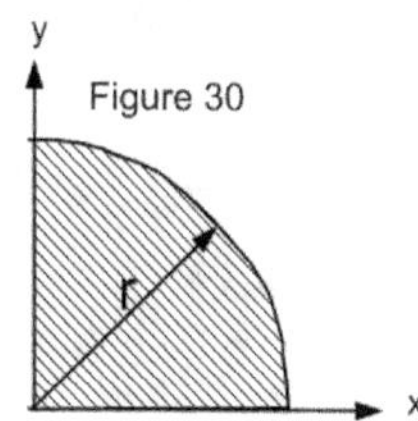

31) Determine the location of centroid of semicircle

Ans: $\bar{x} = 0$; $\bar{y} = \dfrac{4r}{3\pi}$

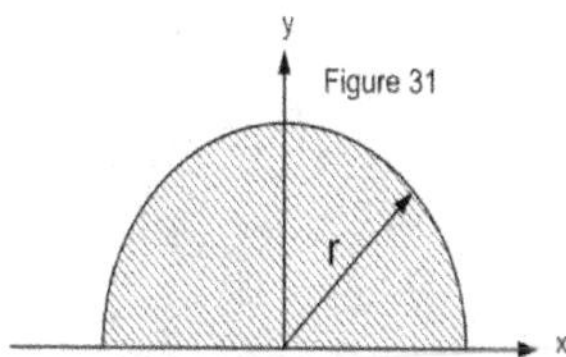

32) Determine the location of centroid of the shadow region

Ans: $\bar{x} = 0.4m$; $\bar{y} = 0.571m$

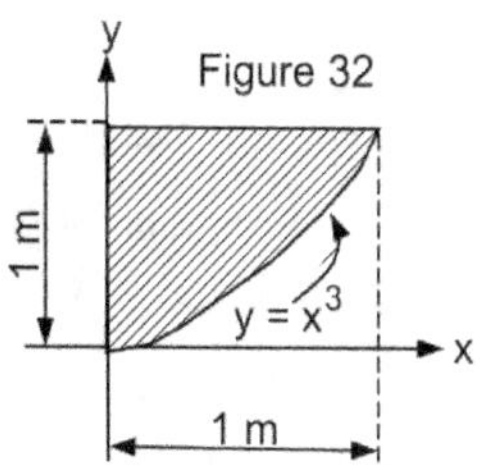

33) Determine the centroid of the shadow region

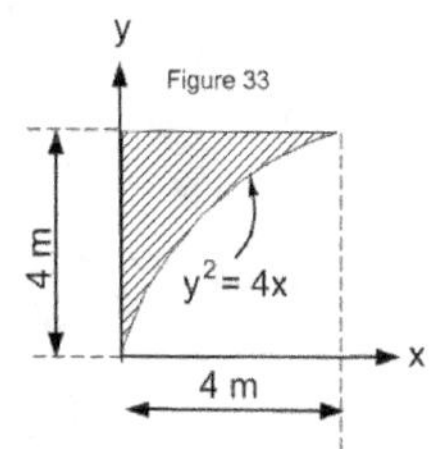

34) Locate the centroid of the plane area shown in figure 1 [- 0.348m; 1.22m]

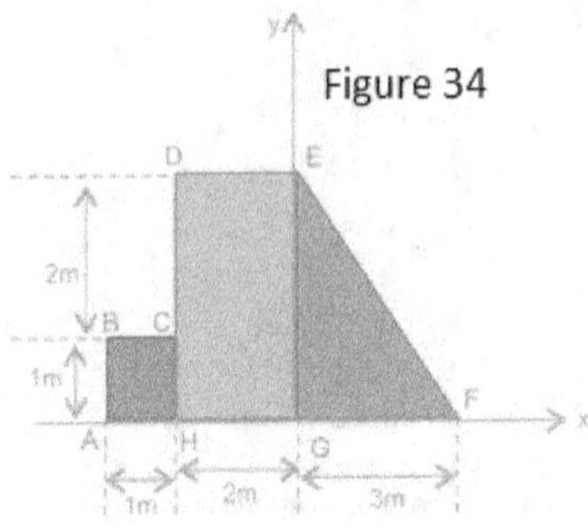

35) Locate the centroid of the plane figure 2. $[\bar{x} = 1.97m = \bar{y}]$

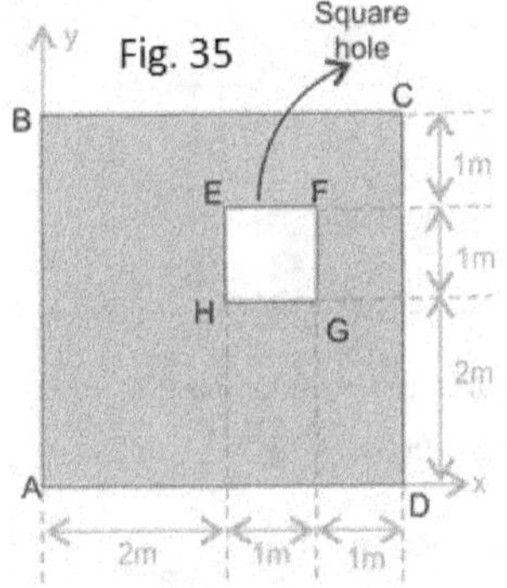

36) Figure 3 shows a thin circular plate with a hole. If the centroid of this plate needs to be located at $\left(\frac{-1}{126}m, \frac{-1}{63}m\right)$, determine the coordinates (x_1, y_1) of the center of hole. [(0.5m, 1m)]

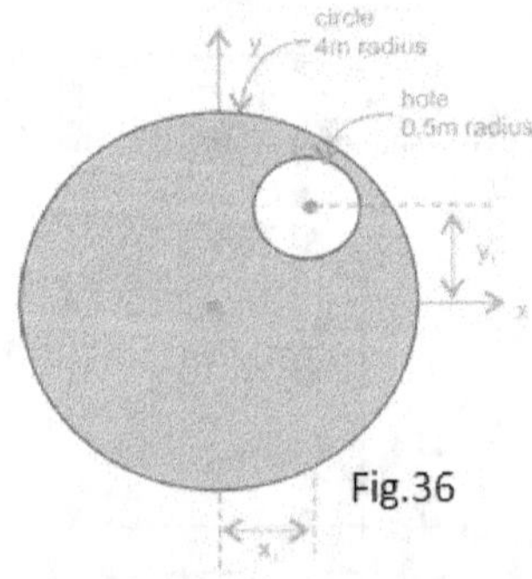

37) Locate the centroid of the curve in the figure 1. [Ans: 0.912m]

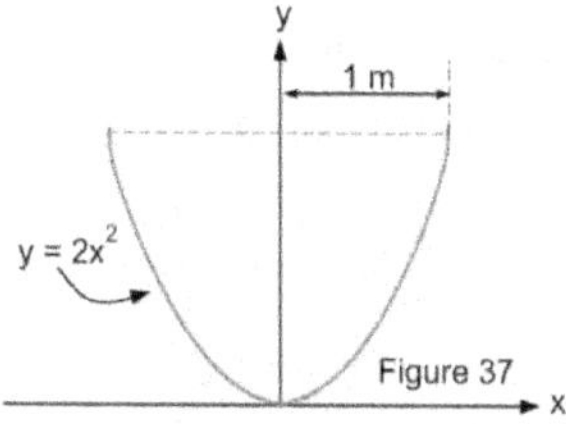

38) Locate the centroid of the homogeneous rod shown in figure 38. If the rod is simply supported at A and hinged at B and has a mass per unit length 0.25 kg/m, determine the reactions at A and B. $\left[\bar{x} = 0.62m; \ A_X = 0 = B_X; \ A_y = 2.1N; \ B_y = 3.54N\right]$

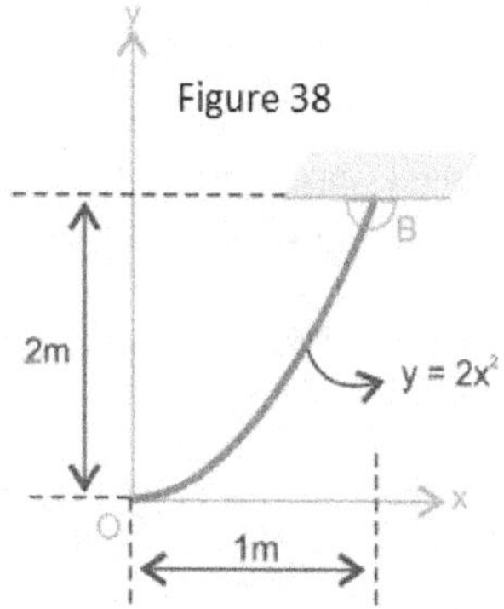

39) Locate the centroid of the curve bent as shown in figure 3. The segment ABC is semicircle with radius = 2m and is lying in xy −plane. The portion CDE is also a semicircle with radius = 2m and is lying in the xz −plane. $\left[\bar{x} = \dfrac{4}{\pi}n; \ \bar{y} = 1m = \bar{z}\right]$

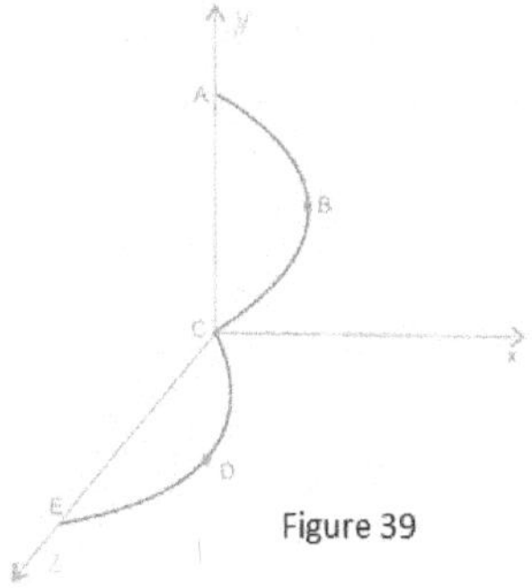

40) A through hole of 15 mm radius is drilled into a composite body as shown in the figure. Determine the location of centroid of this composite body. The cube has an edge of 100 m and the height of pyramid is 90 mm.

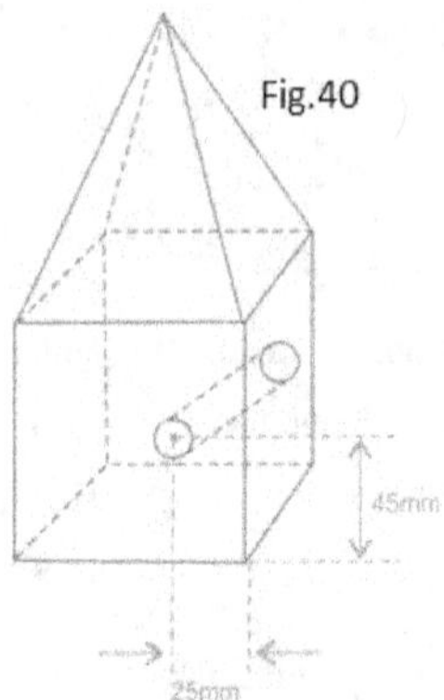

41) A cube portion of 25 mm side is cut and removed from the corner of a cube of 100 mm side as shown in Figure below. Determine the location of centroid of solid thus obtained.

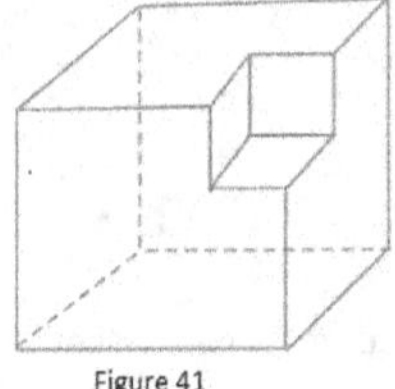

Figure 41

42) Compute the non-flat surface area of a hemisphere using Pappu's Theorem.

43) Compute the surface area of a right circular cylinder using Pappu's Theorem.

44) Sketch 3-D object obtained by rotation of (a) circle and (b) semi-circle as shown in figure 44 and 45 about the axes shown and find their surface areas using Pappu's Theorem.

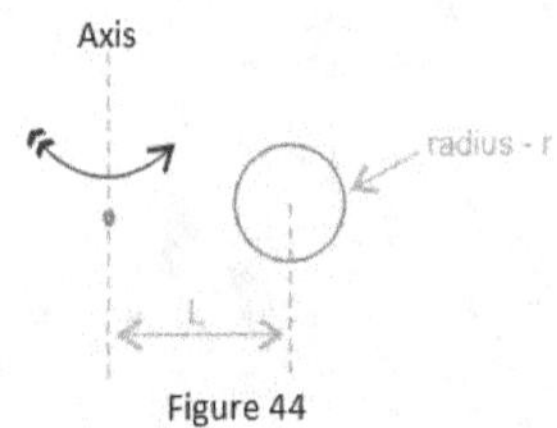

Figure 44

45) Compute the volume of a right circular cylinder of radius "r" and height "h" using theorem of Pappus.

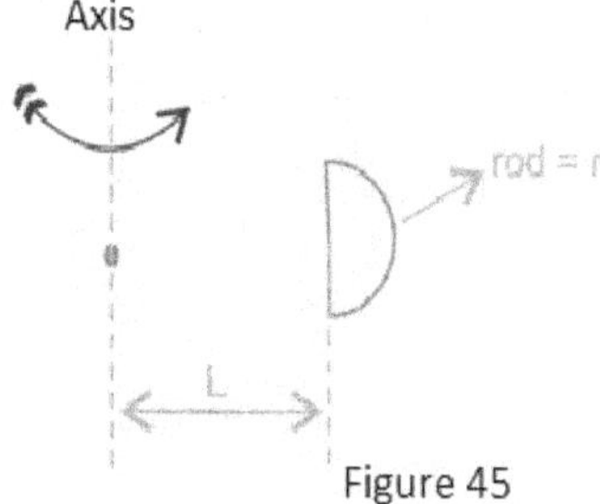

Figure 45

46) Compute the volume of a sphere of radius "r" using Theorem of Pappus.

47) Compute the volume of the object obtained by revolving a circle in Figure 44.

48) Compute the volume of the object obtained by revolving semicircle in Figure 45.

49) Find moments of Inertia of shaded areas about axes shown

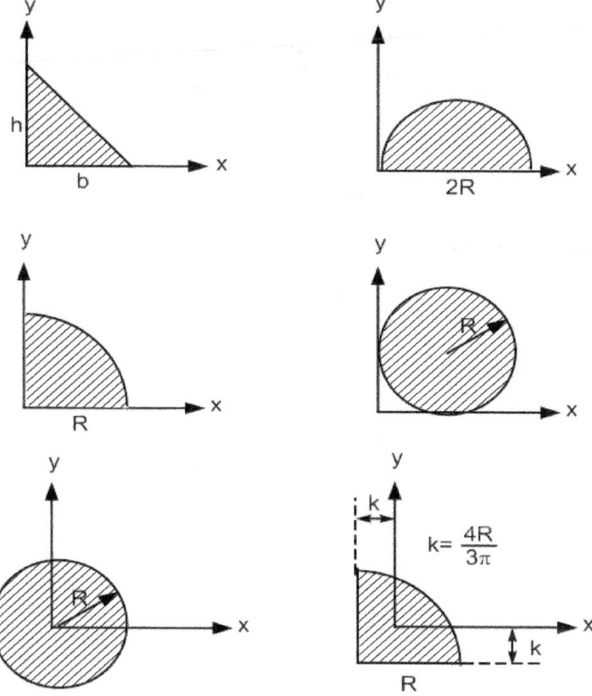

50) Find moment of inertia of the shaded area in figure 50 w.r.t. x −axis

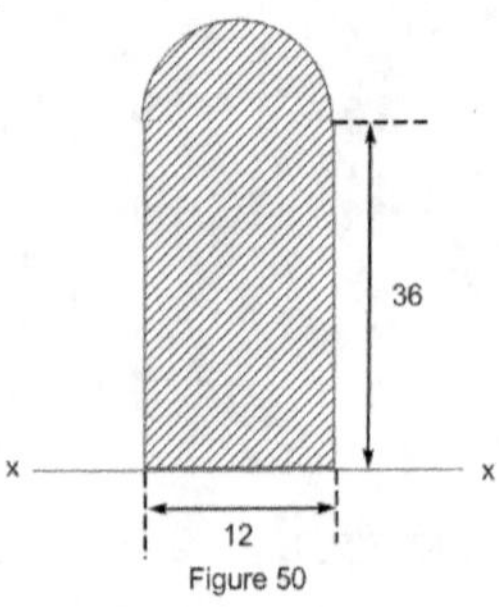

Figure 50

51) Compute moment of inertia of I-section shown in figure 51 w.r.t. x-axis

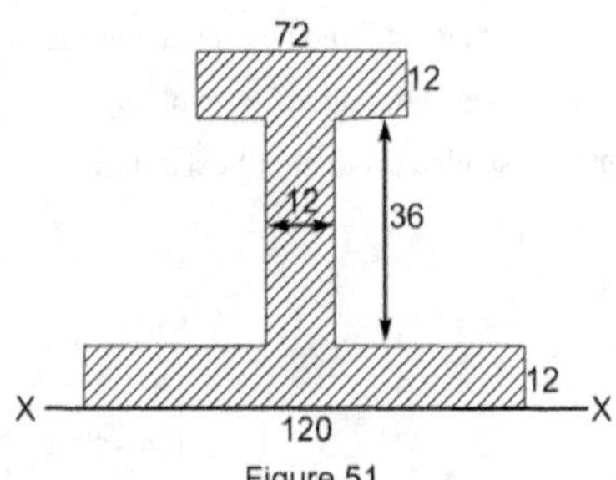

Figure 51

ALL DIMENSIONS ARE IN MM.
SKETCH IS NOT DRAWN TOSCALE.

52) Find the mass moment of inertia of a right circular cone of base radius R, height h, material density P, and a total mass of M, about three perpendicular axes xx, yy, zz as shown in figure 4.

Ans: $\left[I_{mz} = 2I_{mx} = 2I_{my} = \frac{3}{10} MR^2 \right]$

Figure 52

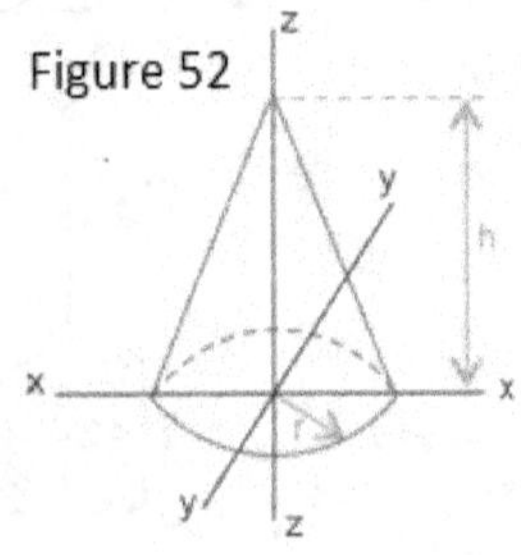

53) Derive expression for Mass moment of inertia of a hollow right circular cylinder of inner radius R_i, outer radius R_o, length L, material density γ, and mass M about $z-z$ axis as shown in figure 5.   Ans: $\left[I_{mz} = M\frac{R_o^2+R_i^2}{2}\right]$

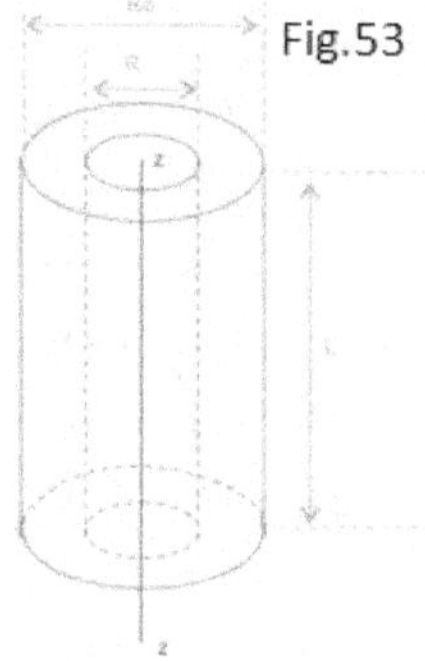

54) Suppose M is the mass of an object of material density ρ. Let G be the center of gravity of the object. Further, let $x-x$ be an axis passing through G. find the value of the volume integral $\int_v r(dm)$ where dm is an element mass and r is the sidtance fro the centroidal axis $x-x$. [Ans: Zero]

55) Derive an expression for mas MI of a solid sphere of radius R and mass M about a centroidal axis. Ans: $\left[\frac{2}{5}MR^2\right]$

56) A car travels in a straight line such that for a short time, its velocity is defined by $\vartheta = (3t^2 + 2t)$ m/s where t is in seconds. Determine its position and acceleration when t = 3sec. initial condition is given as s=0 at t=0. [Ans: s = 36m; a = 20 m/s^2]

57) A sandbag is dropped from a balloon which is ascending vertically at a constant speed of 6m/s. if the bag is released with the same upward velocity of 6m/s when t=0 and hit the ground when t = 85, determine the speed of the bag as it hits the ground and the altitude of the balloon at this instant. [Ans: 72.5 m/s; 314.72 m]

58) An automobile starting from rest increases its speed from 0 to 10 m/s with constant acceleration 1 m/s^2, runs at this speed for a time and finally comes to rest with constant deceleration of 2 m/s^2. If the total distant travelled is 3km find the total time required for travel. [Ans: 307.5 sec]

59) A particle moves along a horizontal path with a velocity of $\vartheta = (3t^2 - 6t)$m/s where t is the time in seconds. If it is initially located at the origin O, determine the distance travelled in 3.5 sec, average velocity and average speed during the interval.
   [Ans: 14.1m; 1.75 m/s; 4.04 m/s]

60) A car approaching a traffic light at 30 m/s sees red and had to stop. If brakes are applied to give a deceleration of 8 m/s² and the car skids to a stop, find the distance over which the car skidded. [Ans: 56.3 m]

61) A stone is thrown vertically up from the ground with a velocity of 91.44 m/s. how long must we wait before dropping a second stone from the top of a 182.88 m tall building if the two stones are to pass each other 60.96 m from the top of the building?
[Ans: 13.67 s]

62) A ball is shot vertically into the air with a velocity of 58.9 m/s. after 4 sec another ball is shot vertically into the air. What initial velocity must the second ball have in order to meet the first ball 117.8 m from ground? [Ans: 48.36 m/s]

63) A projectile is fired up the inclined plane at an initial velocity of 15 m/s. the plane is making an angle of 30° from the horizontal. If the projectile was fired at an angle of 30° from the incline compute the maximum height z measured perpendicular to the incline that is reached by the projectile. Neglect the air resistance. [Ans: 3.31 m]

64) A girl throws a ball with an initial velocity of 10 m/s as shown in figure beside. If time of flight is 5 sec, find height h and vertical component of velocity wen it hits ground.
[Ans: 122.63 m; 49.05 m/s]

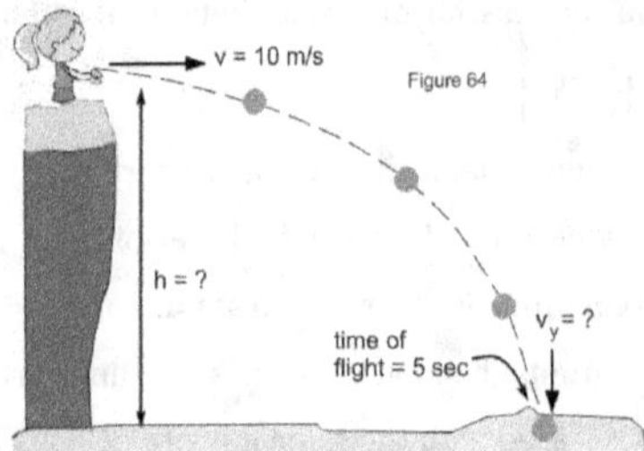

65) When designing a highway curve, it is required that cars travelling at a constant speed of 25 m/s must NOT have an acceleration that exceeds 3 m/s². Determine the minimum radius of curvature of the curve. [Ans: $r \geq 208.3$ m/s]

66) A car travels along the curve having a radius of 300 m. if its speed is uniformly increased from 15 m/s to 27 m/s in 3 sec. determine the magnitude of its acceleration at the instant its speed is 20 m/s. [Ans: 4.22 m/s²]

67) Starting from rest a byclist travels around a horizontal circular path of $\rho = 10m$ at a speed of $\vartheta = (0.09t^2 + 0.1t)$ m/s where t is in seconds. Determine the magnitude of his velocity and acceleration when he has traveled s = 3m starting from initial position.
[Ans: 1.96 m/s; 0.93 m/s²]

68) A truck travels in a circular path having 50 m radius at a speed of 4 m/s. for a short distance from s = 0 its speed is increased by ϑ = 0.05s m/s² where s is in meters. Determine its speed and the magnitude of its acceleration when it has moved s = 10m. [Ans: 4.58 m/s; 0.653 m/s²]

69) The satellite s travels around the earth in a circular path with a constant speed of 20 m/hr. if the acceleration is 2.5 m/s², determine the altitude h. assume that earth's diameter to be 12713 km. [Ans: 5.99 mm]

70) The van in figure is traveling at 20 km/hr when the coupling of the trailer at A fails. If the trailer has a mass of 250 kg and coasts 45 m before coming to rest, determine the constant horizontal force F created by rolling friction which causes the trailer to stop. [Ans: 85.73 N]

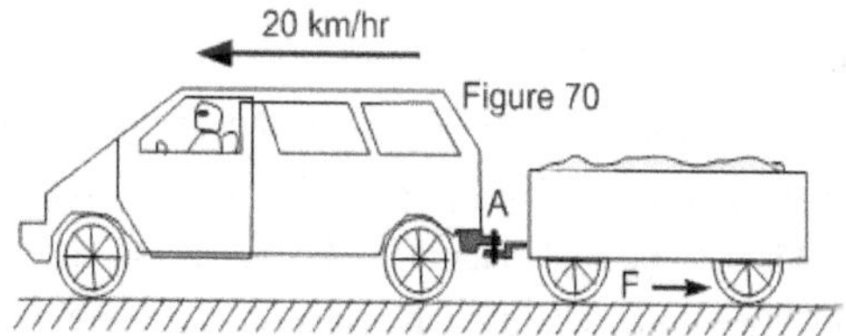

71) A block of 80 kg mass rests on a level ground as shown in figure 71 beside find the magnitude of the force required to give an acceleration of 2.5 m/s² to the block if μ_k = 0.25 between block and ground. [P = 535 N]

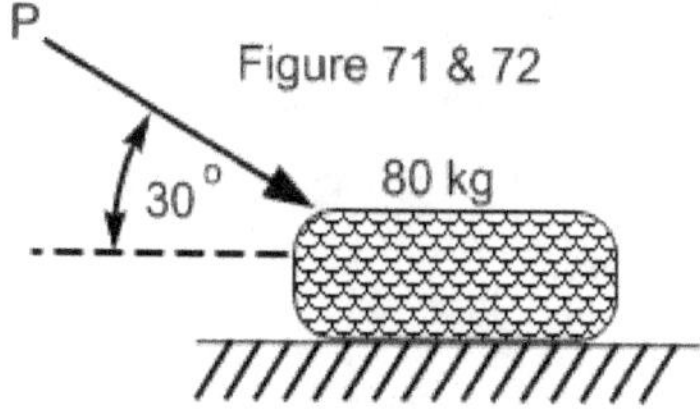

72) The block shown in figure 72 is subjected to a constant force of 298N. If the block, initially at rest, moves over a distance of 12 m in 4 sec, find kinetic coefficient of friction between the block and the ground. [Ans: μ_k = 0.148]

73) Two blocks of masses m_A = 100 kg and m_B = 300 kg are connected as shown in the figure. Ignoring friction at all points of contact, find accelerations of masses and tensions in strings.

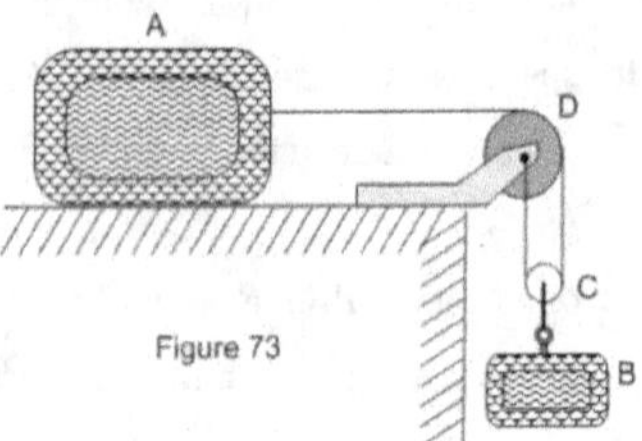

Figure 73

[Ans: a_A = 8.4 m/s²; a_B = 4.2 m/s²; T_1 = 840 N; T_2 = 1680 N]

74) Determine the banking angle θ for the race track so that the wheels of the racing cars will not have to depend upon friction to prevent any car from sliding up or down track. Assume the cars have negligible size, a mass m, and travel around the curve of radius ρ with a speed ϑ. $\left[\tan \theta = \dfrac{\vartheta^2}{\rho g}\right]$

75) A 3 kg disk D is attached to the end of a cord as shown in figure. The other end of the cord is attached to a ball and soket joint located at the center of a platform. If the platform is rotating rapidly, and the disk is placed on it and released from rest as shown, determine the time it takes for the disk to reach a speed great enough to break the cord. The maximum tension the cord can sustain is 100 N and the coefficient of friction between disk and platform is M_k = 0.1. [Ans: 5.89 sec]

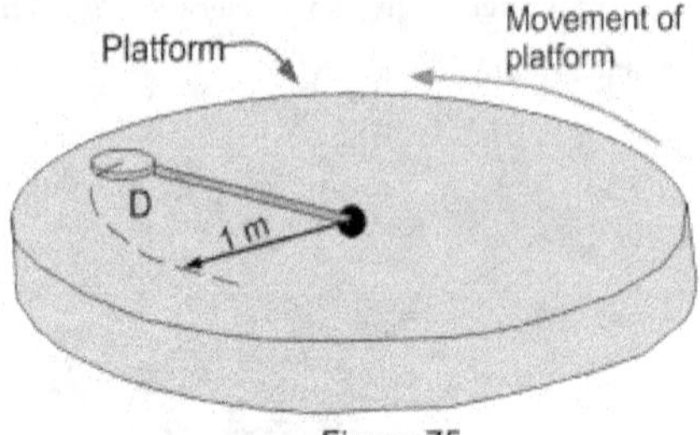

Figure 75

76) When the driver applies the brakes of a light truck travelling at 40 km/h, it skids 3 m before stopping, how far will the truck skid if it is traveling 80 km/hr when the brakes are applied? [12 m]

Figure 76

77) The 100 kg crate is subjected to forces F_1 = 800 N and F_2 = 1.5 KN as shown. If it is originally at rest, determine the distance it slides in order to attain speed of ϑ = 6 m/s. The coefficient of kinetic friction between the crate and the surface is μ_1 = 0.2. [0.933m]

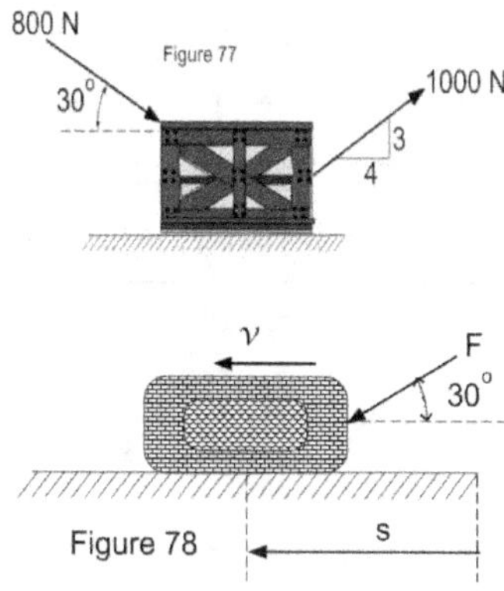

78) The 2 kg block is subjected to a force having a constant direction and a magnitude of F = [300/(1+s)]N where s is in meters. When s = 4 m, the block is moving to the left with a speed of 8 m/s. determine its speed when s= 12 m. The coefficient of kinetic friction between block and the ground is μ_k = 0.25. [15.4 m/s]

1.23. Multiple Answer Questions

1	Is it possible to prove Newton's third law?
A	For Planar Mechanics, it is possible
B	Possible easily
C	For rotation, it is easy
D	There is no "proof".
2	What is the full name of Newton?
A	Sir Isaac Newton
B	Bolton Newton
C	Newton John
D	nobody knows his full name.
3	Select the statement that gives closest meaning to one part of Newton's first law?
A	A body can be in equilibrium whether it is moving or whether it is stationary.
B	A body can be in equilibrium only when it is not moving.
C	If several forces act on a body, the body cannot be in equilibrium.

D	None of the above.
4	How many axioms define the premise of theoretical back ground for engineering mechanics?
A	3
B	5
C	9
D	7
5	Which equation below describes Varignon's Principle closest?
A	$\|Rx\| = \sum \|F_i x_i\|$
B	$\|Rx + \sum F_i x_i\| = 0$
C	$\vec{R}_i x = \sum \vec{F}_i x_i$
D	$Rx = \sum F_i x_i \quad algebraically$
6	Is it possible to prove Varignon's principle?
A	For planar Mechanics problems, yes.
B	For Rotational Mechanics, yes.
C	In some difficult situations, we will not be able to prove.
D	Yes.
7	What do you mean by "Proof of an Engineering Mechanics " Theorm?
A	Checking the algebraic consistency of the theorem to be proven.
B	Use principles of calculus to derive the theorem.
C	Showing that the theorem can be shown to be equivalent to one or more axioms by techniques of proof.
D	A and B.
8	How many equations are sufficient to establish the equilibrium of a Mechanical system of objects under planar condition?
A	4
B	3
C	2
D	A and B
9	Why are counter clock wise moments taken to be positive? Pick the answer that makes sense?

A	Because Newton defined it.
B	Because Einstein defined it.
C	Because fingers don't curl the other way.
D	Because it is convenient.
10	Which three details need to completely specify the effect of force on a body?
A	Direction Cosines, Magnitude, point of application.
B	Moment, direction, Magnitude.
C	Sense of rotation, direction, Magnitude.
D	None of the above completely defines a force.
11	How do you distinguish a body undergoing pure rotary motion?
A	All points on the body move in circles with their own centers.
B	All points on the body move with one point as center.
C	All points on the body move in circles, centers of which lie on a straight line.
D	Very difficult to distinguish
12	Which of the following sentences is most correct about center of gravity?
A	It is a point lying on the body.
B	It may not lie on the body.
C	It does not exist. For convenience we use it.
D	It does not lie on body.
13	Which of the following statements about parallelogram law and triangle law is correct?
A	Triangle law can be deduced from parallelogram law.
B	Parallelogram law can be deduced to Triangle law.
C	Both of them are axioms and cannot be proven.
D	None of the above statements is true.
14	Is Lami's Theorem an axiom? Pick an answer that makes most sense?
A	Yes. Hence it cannot be proven.
B	No. Proving it involves calculus and parallelogram law.
C	No. Proving it involves trigonometry and triangle law.
D	Yes. But it can be proven.
15	If three forces keep an object in equilibrium, which of the following statements can be said to be true about the forces?
A	The forces are coplanar.
B	The forces are concurrent.

C	The forces are coplanar and concurrent.
D	It is not possible to generalize.
16	Which of the following statements is not true about principle of transmissibility?
A	It is not an axiom of engineering Mechanics since it can be proven using trigonometry.
B	It is one of the axioms of Engineering Mechanics.
C	It means that external effect of a force does not change if point of application is moved along its line of action.
D	Internal stresses will vary when force is moved along its line of action.
17	Is it possible to have system of 100 forces which do not have a resultant force? Pick closest answer.
A	Yes it is possible.
B	It is not possible.
C	It depends on the situation.
D	The question is meaningless.
18	Who is one of the authors of one of the prescribed / reference text books for the Mechanics course you are studying?
A	Popov
B	Newton
C	Kalishnikov
D	Young
19	Which of the following statements about our Engineering Mechanics course is true?
A	It involves Mathematics.
B	It involves Common sense.
C	It involves axioms of Mechanics.
D	All above are correct.
20	Will the body in figure beside slide down if coefficient of fraction between the body and the plane is 0.5? (Q.20) W 15°
A	Yes, it will slide down.

B	No, it will not slide down.
C	Need to know the weight of the block.
D	It is NOT possible to decide.
21	Let $\overrightarrow{F_1}$ be the horizontal force that causes movement of the block down the inclined plane and $\overrightarrow{F_2}$ be the horizontal force that causes the block to move up the inclined plane. Which of the following statements is correct? W θ (Q.21)
A	It is not possible to answer question without knowing the values of W and θ.
B	$\overrightarrow{F_2} = \overrightarrow{F_1}$
C	$\overrightarrow{F_2} = -\overrightarrow{F_1}$
D	$|\overrightarrow{F_2}| \geq |\overrightarrow{F_1}|$
22	Is friction a desirable phenomenon? Pick the answer that makes most sense.
A	Yes it is needed very much.
B	No it is best if there is no friction in universe.
C	Sometimes desirable and sometimes not desirable.
D	Dry friction is desirable.
23	Which of the following equations best describes Newton's Second law?
A	$\vec{F} = m\vec{a}$
B	$F = ma$
C	$\vec{F} \propto \dfrac{d}{dt}\{m\vec{v}\}$
D	$\vec{F} = \dfrac{d}{dt}\{m\vec{v}\}$
24	Do you need knowledge of Calculus to solve problems of Engineering Mechanics? Pick the answer that makes most sense.
A	It is essential to solve every problem of Engineering Mechanics.
B	It is NOT at all needed. Just Trigonometry is sufficient.
C	In solving some problems involving maxima and minima, both Calculus and Trigonometry need to be used.
D	None of the above statements make any sense.

25	Is there any axiom of engineering Mechanics that cannot be proven? Pick the answer that makes most use.
A	No axiom of Engineering Mechanics can be "proven." We just have to accept them to be true.
B	There are two axioms that can not be proven.
C	There are three axioms that can not be proven.
D	All axioms of Engineering Mechanics can be proven.
26	What can you say about the resultant force of the system of forces shown in Figure Q.26?
A	The resultant force acting on the body is zero.
B	There are two moments that are trying to rotate the objecct and hence the object is not in equilibrium.
C	There is no resultant force but there is a resultant moment.
D	None of the above statements are true.
27	Consider the L-beam in Figure Q.27. Which of the following statements is not true about the beam?
A	The L-beam is in equilibrium.
B	The net moment on the beam is -P(a+b).
C	There is no single resultant force for this system of forces.
D	The beam will rotate in colock-wise direction under action of forces shown.
28	Two forces $\vec{P}$ and $\vec{Q}$ are acting on the I-beam as shown in Figure 28. A third force, $\vec{R}$ when applied on the beam, keeps it in equilibrium. Pick the statement that most correctly applies to the situation.

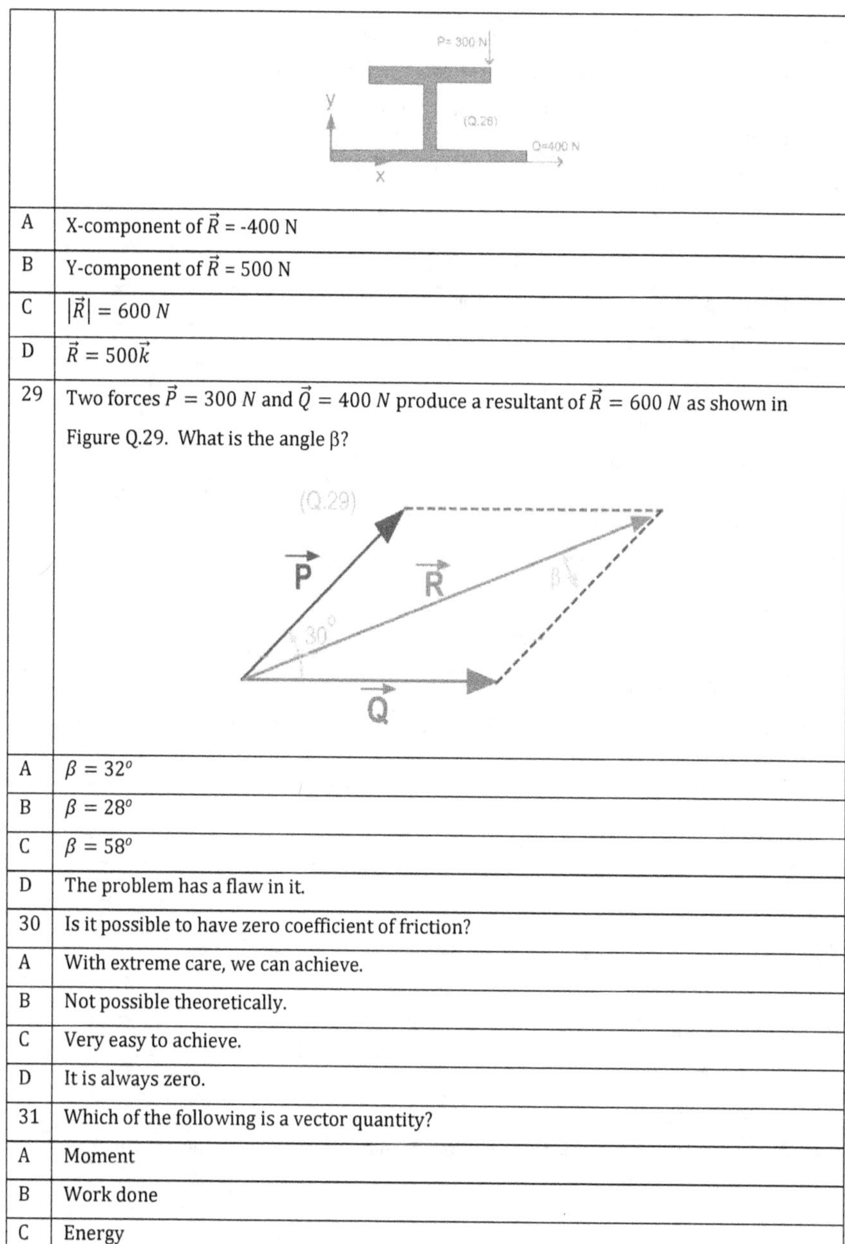

A	X-component of $\vec{R}$ = -400 N
B	Y-component of $\vec{R}$ = 500 N
C	$\|\vec{R}\| = 600\ N$
D	$\vec{R} = 500\vec{k}$
29	Two forces $\vec{P} = 300\ N$ and $\vec{Q} = 400\ N$ produce a resultant of $\vec{R} = 600\ N$ as shown in Figure Q.29. What is the angle β?

A	$\beta = 32^o$
B	$\beta = 28^o$
C	$\beta = 58^o$
D	The problem has a flaw in it.
30	Is it possible to have zero coefficient of friction?
A	With extreme care, we can achieve.
B	Not possible theoretically.
C	Very easy to achieve.
D	It is always zero.
31	Which of the following is a vector quantity?
A	Moment
B	Work done
C	Energy
D	Speed

32	A body of weight W is required to move up on rough inclined plane whose angle of inclination with the horizontal is α. The effort applied parallel to the plane is given by (where $\mu = \tan\varphi$ = Coefficient of friction between the plane and the body.)
A	$P = W \tan(\alpha + \varphi)$
B	$P = W (\cos\alpha + \mu\sin\alpha)$
C	$P = W \tan\alpha$
D	$P = W (\sin\alpha + \mu\cos\alpha)$
33	A couple produces which type of movement?
A	Pure rotary motion.
B	Pure translatory motion.
C	Translatory + Rotary Motion
D	Keeps the body in equilibrium.
34	The angle of inclination of the plane when the body begins to move down the plane is called
A	Angle of Repose
B	Angle of Friction
C	Angle of Sliding down
D	Angle of Projection
35	The resultant of the two forces P and Q is R. If Q is doubled, the new resultant is perpendicular to P. Then
A	P=Q
B	Q=R
C	P=R
D	None of the above
36	Friction experinced by a body when it is at rest is called:
A	Dynamic friction
B	Static Friction
C	Limiting Friction
D	Coefficient of friction
37	A ladder is resting on a smooth ground and leaning against a rough vertical wall. The force of friction will act
A	downward at its upper end
B	towards the wall at its upper and

C	away from the wall at its upper end
D	upward at its upper end
38	A number of forces meeting at a point will be in equilibrium when
A	sum of all the forces is zero
B	all the forces are equally inclined
C	Sum of resolved parts is zero (i.e. $\sum V = 0$ and $\sum H = 0$)
D	None of the above
39	The forces, whose lines of action are parallel to each other and act in the same directions, are known as
A	Coplaner, Concurrent forces
B	Like Parallel forces
C	Coplaner, non-Concurrent forces
D	Unlike, Parallel forces
40	Two forces are acting at an angle of 120°. The bigger force is 40N and the resultant is perpendicular to the smaller one. The smaller force is
A	10 N
B	20 N
C	30 N
D	35 N
41	Is area moment of inertia a scalar quantity?
A	Yes, it is always a scalar quantity.
B	No, it is always a vector quantity.
C	Sometimes a scalar quantity.
D	It is not possible to define.
42	Who proposed theorems of Pappus?
A	Pappus
B	Guldinus
C	Pappus and Guldinus
D	None of the above
43	Under which type of situations does Work-Energy principle find its application very convenient?
A	When force, velocity, and time of movement are known.
B	When force, coefficient of friction, and distance are known.

C	When distance, force, and velocity are known.
D	None of the above.
44	Under which type of situations does Impulse-Momentum principle find its application very convenient?
A	When force, coefficient of friction, and distance are known.
B	When distance, force, and velocity are known.
C	When force, velocity, and time of movement are known.
D	When force, coefficient of friction, and angle are known.
45	For a particle moving in a circular path of radius "r" with constant speed, which of the following statements is true?
A	Its normal component of acceleration is constant.
B	Magnitude of its tangential component of velocity is not constant.
C	Vector quantity representing its total acceleration is constant.
D	Magnitude of normal component of acceleration is constant.
46	Is it possible to prove Work-Energy principle?
A	In static situations, it can be proven.
B	In Linear movement situations, it can be proven.
C	In some difficult situations, we will not be able to prove.
D	Yes, it is always possible to provide a proof.
47	Is it possible to prove Impulse-Momentum Principle?
A	In static situations, it can be proven.
B	In Linear movement situations, it can be proven.
C	In some difficult situations, we will not be able to prove.
D	Yes, it is always possible to provide a proof.
48	Is angular momentum a vector quantity?
A	When the body is rotating with uniform velocity, it is vector quantity.
B	When the body is moving in a straight line, it is a vector quantity.
C	Sometimes it is a vector quantity and sometimes, it is a scalar quantity.
D	It is always a vector quantity.
49	Is center of gravity a point on the body?
A	No, it is not a point on the body.
B	Yes, it is a point on the body.
C	Sometimes it is a point and sometimes it is not.
D	For convenience, it is defines. In actual practice, it does not exist.

50	Where do you locate the center of gravity of a hollow sphere of inner radius R_1 and outer radius R_2?
A	At topmost point on the inner surface of the sphere.
B	At bottommost point on the outer surface of the sphere.
C	At the bottommost point on the inner surface of the sphere.
D	At the center of the outer surface of the sphere.
51	A solid body is rotating about an axis passing through its centre of gravity with an angular velocity of $\vec{\omega}$. Then, the linear velocity of a point which is at a radial distance of $\vec{r}$ from the axis is given by the expression:
A	$\vec{r} \times \vec{\omega}$
B	$\vec{\omega} \times \vec{r}$
C	$\vec{r} \times (\vec{\omega} \times \vec{r})$
D	$\vec{\omega} \times (\vec{\omega} \times \vec{r})$
52	Vector expression for normal component of acceleration is given by:
A	$\vec{r} \times \vec{\omega}$
B	$\vec{\omega} \times \vec{r}$
C	$\vec{r} \times (\vec{\omega} \times \vec{r})$
D	$\vec{\omega} \times (\vec{\omega} \times \vec{r})$
53	Work done by the force P = 10 N in the figure in moving the block through a distance of 10 cm on the floor is:
A	It cannot be calculated.
B	It is imaginary.
C	$\cos 30^\circ$ N-m
D	None of the above.
54	If a body starts from rest at x=0 and starts moving along the x-axis with an acceleration of $a = 1.5x^2$, what is its velocity when it has traveled a distance of x=100 m?
A	$\sqrt[3]{100^2}$ m/s
B	$\sqrt[2]{100^3}$ m/s
C	$\sqrt[2]{100^5}$ m/s
D	$\sqrt[3]{100^5}$ m/s

55	Coriolis component comes into picture when:
A	A body is traveling along a line with non-uniform acceleration.
B	A body is rotating with non-uniform angular velocity.
C	When a body is rotating about an axis which itself is rotating.
D	When a body is moving along a straight line which is rotating about an axis.
56	Mass moment of inertia of a right circular cone of radius "R", height "h", and mass "M" about a base diametral axis is:
A	$\dfrac{3}{20}MR^2$
B	$\dfrac{3}{10}MR^2h$
C	$\dfrac{3}{20}MR^2h$
D	$\dfrac{3}{21}MR^2h$
57	Mass moment of inertia of a hollow cylinder of inner diameter R_i, outer diameter R_o, length L, material density γ, and mass M about the line joining the end circles is:
A	$M\dfrac{R_i^2 + R_o^2}{2}L$
B	$M\dfrac{R_i^2 + R_o^2}{2}$
C	$\gamma\dfrac{R_i^2 + R_o^2}{2}L$
D	$\gamma\dfrac{R_i^2 + R_o^2}{3}L$
58	Suppose M is the mass of an object of material density ρ. Let G be the center of gravity of the object. Further, let X-X be an axis passing through G. Then, the value of the integral $\int r \cdot dm$ evaluated over the entire volume of the object where dm is the elemental mass and r is the distance of the CG of the elemental mass from X-X is:
A	It is not possible to evaluate the integral
B	It is possible to evaluate but it depends on the shape of the object
C	Zero
D	None of the above
59	The mass moment of inertia of a sphere of radius R and mass M about a centroidal axis is:
A	Depends on the density of the material of the sphere

B	$\dfrac{4}{3}M\pi R^3$
C	$\dfrac{2}{5}MR^3$
D	$\dfrac{2}{5}MR^2$
60	A car travels in a straight line such that for a short time, its velocity is defined by $v = (3t^2 + 2t)\ m/s$ where t is in seconds. Then, its position, and acceleration at t=3 sec with initial conditions t=0, s=0 may be given as:
A	s=26 m, a = 20 m/s²
B	s=36 m, a = 20 m/s²
C	s=36 m, a = 26 m/s²
D	s=30 m, a = 20 m/s²
61	Just after the fan in Figure is turned on, the motor gives the blade an angular acceleration of $\alpha = 20e^{-0.6t}$ rad/s where t is in seconds. How many revolutions the fan blades turn within 3 seconds?
A	2 revolutions
B	25 *revolutions*
C	Not possible to compute.
D	8.5 *revolutions*
62	Cars with mass M and velocity v travel along a curved path of radius r. The banking angle θ such that the cars do not have to depend on friction to prevent any car from slipping up or down along the track is given by:
A	$\tan^{-1}\left(\dfrac{mv^2}{g}\right)$
B	$\tan^{-1}\left(\dfrac{mv^2}{rg}\right)$
C	$\tan^{-1}\left(\dfrac{v^2}{rg}\right)$
D	None of the above
63	A satellite travels around earth in a circular path with a constant speed of 20 Mm/hr. If the acceleration of the satellite is 2.5 m/s², determine the altitude of the satellite. Assume that the radius of earth is 12713 km.
A	5.99 Mm
B	8.99 Mm

C	2.99 Mm
D	Not possible to compute with given data
64	The volume of a torus of mean radius R, generated by a circle of radius r is:
A	$2\pi r^2 R$
B	$2\pi r^2 R^2$
C	$2\pi^2 r^2 R$
D	It is not possible to compute the volume using given data.
65	The surface area of a torus of mean radius R, generated by a circle of radius r is:
A	$4\pi^2 Rr$
B	$4\pi^2 R^2 r$
C	$4\pi Rr^2$
D	$4\pi^2 r^2$
66	Is mass moment of inertia a scalar quantity?
A	Yes if the object is rotating with constant velocity.
B	Yes if the object is rotating with constant acceleration.
C	Yes. It doesn't have a direction associated with it.
D	None of the above.
67	Which of the following expressions represent the net impulse of a force acting over a time period from zero to "t"?
A	$\int_0^t F(\tau)\, d\tau$
B	$\int_0^t \lvert F(\tau)\rvert\, d\tau$
C	$\int_0^t F(t)\, dt$
D	$\int_0^t \lvert F(t)\rvert\, dt$

68	Is the impulse integral given by the expression $J = \int_0^t F(\tau)\, d\tau$ a scalar quantity?
A	No. J is a vector quantity.
B	Only in rectilinear motion J is vector.
C	Only in rotational motion J is vector.
D	J is always a scalar quantity.
69	Which of the following equations indicate the impulse-momentum principal in mathematical form? Choose the closest and most accurate one.
A	$$F \cdot \Delta t = mv_1 - mv_2$$
B	$$\int_0^t \vec{F}(\tau)d\tau = \int_0^t m\, d\vec{v} + \int_0^t \vec{v}\, dm$$
C	$$F \cdot \Delta t = m_1 v_1 - m_2 v_2$$
D	$$F \cdot \Delta t = m_2 v_2 - m_1 v_1$$
70	Is radius of gyration a vector quantity?
A	Yes. It is a vector quantity.
B	In some situations, it is a vector quantity and some situations it is a scalar quantity.
C	It is a scalar quantity.
D	None of the above is answer to the question.
71	Which of the statements below describes the physical significance of the radius of gyration as applied to rotational dynamics? Pick the one closest in meaning.
A	The radius of gyration is the "equivalent distance" of the mass from the axis of rotation.
B	The radius of gyration is the ratio of area moment of inertia to the mass.
C	The radius of gyration is the square root of ratio of area moment of inertia to the mass.
D	None of the above statements give the phsical significance of radius of gyration.
72	What is the formula for escape velocity with usual notations?
A	$$2 \cdot \sqrt{\dfrac{GM}{R}}$$
B	$$\dfrac{2\sqrt{GM}}{R}$$
C	$$\sqrt{\dfrac{2GM}{R}}$$
D	$$\dfrac{\sqrt{2GM}}{R}$$

73	The coordinates of vertices of a triangle are (10,12), (15,25), and (20,23). The coordinates of the centroid of the triangle are:
A	(20,15)
B	(10,15)
C	(15,10)
D	(15,20)
74	The coordinates of end points of diameter of a semicircle are $(-8\pi,0)$ and $(4\pi,0)$. The centroid of this semicircle will be at:
A	(2,8)
B	(4,6)
C	(0,8)
D	(8,0)
75	The coordinates of end points of diameter of a semicircle are $(-8\pi,0)$ and $(4\pi,0)$. The centroid of this semicircle will be at:
A	(2,8)
B	(4,6)
C	(0,8)
D	(8,0)
76	The coordinates of cetroid of a wire bent in the form of a triangle shown in the figure are: 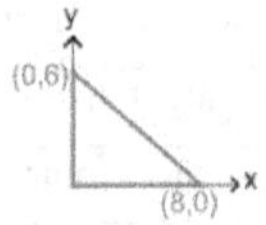

A	(2,3)
B	(3,2)
C	(4,3)
D	(3,4)
77	A wire is bent into the form of a semicircular portion and a straight portion as shown in the figure. The coordinates of centroid of this bent wire are: 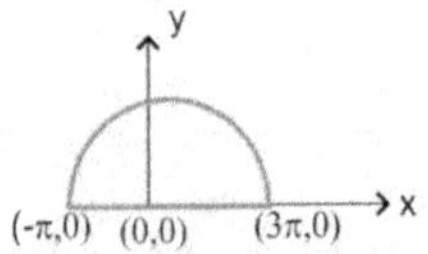

A	(0,4)
B	$\left(\pi, \dfrac{\pi}{\pi + 2}\right)$
C	$(\pi,4)$
D	$(4,\pi)$
78	The center of gravity of a square pyramid with base ABCD with side a, base center at O, apex at P, and height OP of h is located at a distance of xh from O on the line OP which is perpendicular to the base ABCE. The value of x is:
A	$\dfrac{1}{4}$
B	$\dfrac{3}{4}$
C	$\dfrac{1}{3}$
D	$\dfrac{2}{3}$
79	The height of center of gravity of a hemisphere of radius r from the flat circular face is:
A	$3r/8$
B	$5r/8$
C	$7r/8$
D	None of the above
80	The mass moment of inertia of a hemisphere of mass m and radius r about a diametral axis of the flat circular face is:
A	$\dfrac{1}{5}mr^2$
B	$\dfrac{4}{5}mr^2$
C	$\dfrac{2}{3}mr^3$
D	$\dfrac{2}{5}mr^2$

www.ingramcontent.com/pod-product-compliance
Lightning Source LLC
Chambersburg PA
CBHW071958150726
47999CB00001B/472